A *New York Times* Notable

USA Today's #1 Nonfiction Book of 2022

Vulture's #1 Memoir of 2022

A *Washington Post*, *Los Angeles Times*, *TIME*, *BuzzFeed*,

Lit Hub, *Publishers Weekly*, *Booklist*, Electric Literature, and

New York Public Library Best Book of the Year

A 2022 Goodreads Choice Award Nominee

An Indie Next Pick

"A rich, decadent book that rewards close reading . . . Anyone who immerses themselves in Chloé's writing will come away with a greater understanding of everything beautiful about the human experience, and how to behold it."

—Isaac Fitzgerald, the *Today* show

"Gorgeous, vivid. . . . As she rejects the dismissive gaze of others, Jones shows how she stands in the light of her own extremely able self."

—*The New York Times*

"[A work] of genuine genius . . . *Easy Beauty* advertises itself as a memoir, which it is, but it's also several other things: a travelog; a philosophical tract; a meditation on what it is to live in this world with a visible disability; a consideration of what beauty means, what parenthood means, what art and fame mean. If this sounds like too much, it's not: Ms. Jones is a brilliant writer, a composer of stunning sentences that nail down experiences and emotions I would have thought ineffable."

—Elizabeth McCracken, *The Wall Street Journal*

"Chloé Cooper Jones' prowess as a writer seemingly knows no bounds. . . . In this insightful, complex, engrossing, and deeply personal book, Jones asks us to reflect on what, and who, we think is beautiful. I've spent all year recommending it to everyone I know."

—Heather Radke, *Esquire*

"Chloé Cooper Jones's debut shifted my understanding of a world I've experienced only while able-bodied. . . . Extraordinary."

—Arianna Rebolini, *Vulture*

"Soul-stretching, breathtaking . . . A profound, impressive, and wiser-than-wise contemplation of the way Jones is viewed by others, her own collusion in those views, and whether any of this can be shifted . . . A game-changing gift to readers."

—*Booklist* (starred review)

"Chloé Cooper Jones is ruthless in probing our weakest and darkest areas, and does so with grace, humor, and, ultimately, with something one seldom finds: kindness and humanity."

—André Aciman,
author of *Call Me by Your Name*

"In this ambitious and elegant book about seeing and being seen, Chloé Cooper Jones invokes thorny, theoretical material about identity, the social order, and how we measure human value, but her clarity and compassion invite all readers in. She has created a forceful and fresh point of view from which to anatomize power, access, and perception in her precise, unsparing prose. A necessary, relentlessly honest book that feels both of the moment and timeless."

—Whiting Foundation Judges' citation

"Transcendent . . . In keeping the reader close as she navigates the world, Jones lets us in on the effort it takes to move through the world in a disabled body. . . . This is all rendered in sentences, insights, and metaphors so precise and evocative that demonstrate her literary mastery."

—*Oprah Daily*

"Jones resists sentimentality and is as unsparing on herself as she is on other people, yet she writes with such graciousness, too. A wonderful debut."

—*Buzzfeed*

"Exceedingly intelligent . . . Jones is the kind of writer who defies boxes."

—Electric Literature

"[In her] dazzling debut . . . Chloé Cooper Jones challenges society's rules of attraction with razor-sharp wit and intellect . . . [and] makes a brilliant case for the beauty of complexity."

—*Publishers Weekly* (starred review)

"Philosophy, art, gender, sex, travel, motherhood, academia, humor—this book has it all."

—New York Public Library

"Written with the curiosity of a scholar, the compassion of a mother, and the keen insight of a person who has lived on the margins of what is deemed acceptable, *Easy Beauty* is a rare, poignant gem of a memoir. . . . Transcendent."

—*Bookreporter*

"Jones challenges the unspoken social taboos about the disabled body, unpacking myths of beauty and our complicity in upholding those myths. Blending journalism, philosophy, and memoir, it's a book that everyone will be talking about."

—*Literary Hub*

"*Easy Beauty* sinks its teeth into you from the first sentence; Jones's razor-sharp prose is drenched in honesty and intellect. [A] necessary—funny, brilliant, moving, and unflinching—read."

—Apartment Therapy

"Jones's writing is thoughtful and deeply felt, and her stories will fascinate anyone who wants to look at the world in a new way."

—Apple Books (Best of the Month)

"Masterful . . . This book will take your breath away."

—*Debutiful*

"[*Easy Beauty*] is life-altering for both those of us who have and those who have not experienced chronic pain. Jones fearlessly invites readers to question what we accept, what we allow, and what we can endure as women in a world that is largely designed to serve able-bodied men."

—*Shondaland*

"I recommend *Easy Beauty* to anyone who has wanted beauty badly, even without knowing quite what it is, but who could never seem to access it. At least, I'm that sort of anyone, and I could feel and recognize parts of myself in every moment of this book. Chloé Cooper Jones's writing pierces right through and lets a light in."

—Mitski

"Perceptive, stylish, and darkly funny, *Easy Beauty* is an act of grace, and a reckoning. Chloé Cooper Jones is a remarkable writer—I would follow her mind anywhere."

—Anna Wiener,
author of *Uncanny Valley*

"Chloé Cooper Jones is a writer whose work I don't read, but enter: she weaves her brainy, crackling interior into the sinews of a reality that is forever reminding its participants of the difficulty of living inside a body. *Easy Beauty* is the most humane book I have read in a long time: in her insistence that we bear witness to each other, Jones calls forth a better, and indeed more beautiful world. I loved this book."

—Kristen Radtke, author of
Seek You: A Journey Through American Loneliness

"Jones is a magnificent guide, fiercely sharp and fiercely human. This book is for anyone who wants to immerse themselves into a world of beauty, who wants to get real about the roots of their desire, and who can't quite kick the habit of admiring the structures—and humans—who harm them. The questions she raises will resound in your head for a long time to come."

—Lulu Miller, author of
Why Fish Don't Exist and cohost of *Radiolab*

"This book is utterly remarkable. I was spellbound by the style, the ideas, the vulnerability, the talent."

—Lydia Kiesling,
author of *The Golden State*

"What a gift of a book . . . *Easy Beauty* has the rigor and precision of Joan Didion and Maggie Nelson and a forthright humor and naked truth all its own."

—Sarah Ruhl, author of *Smile*

EASY BEAUTY

a memoir

CHLOÉ COOPER JONES

Avid Reader Press

NEW YORK · LONDON · TORONTO · SYDNEY · NEW DELHI

Avid Reader Press
An Imprint of Simon & Schuster, LLC
1230 Avenue of the Americas
New York, NY 10020

Names and identifying characteristics of some individuals have been changed and some dialogue has been re-created.

First Avid Reader Press trade paperback edition April 2023

Avid Reader Press and colophon are trademarks of Simon & Schuster, LLC

Interior design by Paul Dippolito

Manufactured in the United States of America

3 5 7 9 10 8 6 4

Library of Congress Control Number: 2022952056

ISBN 978-1-9821-5199-7
ISBN 978-1-9821-5200-0 (pbk)
ISBN 978-1-9821-5201-7 (ebook)

For Yoyo

The psyche is a historically determined individual relentlessly looking after itself. . . . The area of its vaunted freedom of choice is not usually very great. One of its main pastimes is daydreaming. It is reluctant to face unpleasant realities. Its consciousness is not normally a transparent glass through which it views the world, but a cloud of more or less fantastic reverie designed to protect the psyche from pain. It constantly seeks consolation, either through imagined inflation of self or through fictions of a theological nature. Even its loving is more often than not an assertion of self. I think we can probably recognize ourselves in this rather depressing description. . . . I want now to return to the beginning and look again at the powerful energy system of the self-defensive psyche in the light of the question, How can we make ourselves better?

—Iris Murdoch, *The Sovereignty of the Good*

EASY BEAUTY

PROLOGUE

The Neutral Room

I am in a bar in Brooklyn listening as two men, my friends, discuss whether or not my life is worth living. Jay is to my left and Colin to my right. Colin, an ethical philosopher trained in my same doctoral program, argues a vision for a better society, one where a body like mine would not exist. The men debate this theory, speaking through me. This is common, both the argument and the way I'm forgotten in it.

The window in front of me frames scenes from the street. Groups of people, unified in exuberant movement, pass by like rowers on a river, propelling themselves swiftly into their Friday night. I wish for one person to stop and meet my gaze, wave me up from my seat and out to the sidewalk, inviting me to follow them into a more fun future. None do.

I don't want to be with these men, at this bar, anymore. I think to fake a phone call, to fill my vacant face with false concern, then walk out, slip into the stream of people, disappear. I am not so far from home. I imagine myself already there, leaning to kiss the forehead of my sleeping son, collapsing in my own bed, drawing my hand across my husband's shoulders. But habit and exhaustion limit me. I am humiliated again.

To speak up, against Colin, would require an energy I don't have and don't want to access, not now, not tonight, not after I've taught classes all day at school, made dinner at home, read the same book—*Cars Go Vroom!*—four times to my five-year-old, brushed my teeth, fixed my hair, put on my favorite dress.

It is early May. A month of April rain filtered the atmosphere and the scent of clean spring air, acidic and sweet, reaches me through the bar's open window. I want to enjoy this warm evening, and I want to think a little less. Simple pleasure is inaccessible to me now, but I know I can have something like it if I stay quiet, let the men talk. I can sit through their conversation from a remove. It won't last forever. So, I seal myself up, I become a statue, I lean against the wall, am bordered by neon light; I try on, keep on, a fixed expression; I leave the scene and become a surface only. The men bicker over the issue of my unfortunate birth. I search for anger and find only numbness.

I center myself in The Neutral Room, a separated space inside my mind I constructed when I was very young as a method for dissociating from physical pain. There are no doors or windows in the neutral room, nothing but white walls, and on the walls, one at a time, gray numbers flash . . .

1 2 3 *4*, 5 6 7 *8*, 1 2 3 *4*, 5 6 7 *8*

. . . and I count them until everything else fades and I'm lost in a void that nullifies, dulls what needs dulling, and from there the world blurs and the bar is both darker and louder, blunting Colin's face and voice, and his words reach me weakly, then dissolve, subsumed below the shriek and grind of bar and street noise, which whirrs on and on, black sound blackening the lasting night.

Part One: A Window

1

The Berninis

Three months later

A stranger is staring at me. I drop my shoulder, glimpse him over it. He is tall. He crosses the room, moving toward me with a long stride, smooth and sure. The stranger's stare fastens, binds me tighter to him as he moves closer. His eyes scrape across my body, then he looks away, back, away, then skips discretion and takes in my length, eyes prowling up and down. Newness incites the eye and I am always a new thing. Once accustomed, he turns from me and looks at the Bernini sculpture in front of us, a scene from Ovid's *Metamorphoses*. Now it is my turn to stare.

The stranger is built by blueprint and ruler. Jaw to neck, shoulder to torso, hip to knee: a body of straight lines, design, intention. I'm flushed, too warm, I stink, sweat drips, rivulets of self-reproach; he is near me now, dry and smiling; it unnerves me, his dryness. I'd left my hotel at noon: an error. Heat made the streets shimmer. The air was sticky and humid as a mouth. Dust, rising in fine mists, drifted over me, left me gritty.

Behind us, the Galleria Borghese bursts with tourists; they push in close, pen us in, making a frame around me and the stranger. From a distance we might look like polite people appreciating a famous sculpture, but from where I stand, inside the mass, I can see all the sly, slow glances, flushed faces, dilations, smiles, pulses, and swells, and I am caught in their undertone, washed by their waves of red energy. Our eyes hang on the sculpture at a single juncture, where Pluto's hand presses deep into Proserpine's naked leg.

The sculpture depicts a story from Roman mythology. One version goes like this: Pluto offends Venus, the goddess of love. As an act of revenge, she tells Cupid to send his arrow through Pluto's heart, afflicting him instantly with a love-like madness. Proserpine, the daughter of the goddess Ceres, is nearby picking flowers. Pluto, god of the Underworld, abducts her, forcing her away from nature and toward the safety of the dark and isolated world he rules.

Bernini stills, for our consideration, the moment when Pluto sees Proserpine and takes her, holds her roughly. He wraps a hard hand around her thigh, and at that point of contact Bernini has made metamorphic rock soft, impossibly. The way marble fingers sink into marble flesh, the eroticism of this aggression—it makes me uneasy, but I don't look away and neither does anyone else.

The stranger inches closer. His elbow finds my shoulder and stays. Where we touch becomes a whole sensate world made of heat, weight, a scent like wet leaves. Then his arm parts from mine, just barely, and the world expands to that narrow space that separates us, and through that space the possibility of adventure trembles forth. Fine hairs and ridged red flesh rise to bridge the gap between my body and his. My thoughts crawl along my skin. The stranger and I take breaths in unison, suspended in anticipation of the other's gesture. I imagine the stranger grasping me as Pluto grasps Proserpine. He leans closer and a budding warmth in me blossoms. A thought toward pleasure: to see him kneel and lick Rome's dust from my bare leg. Just then the stranger tips forward and inhales sharply as if this would dislodge a tiny particle of the Bernini that he could ingest, something to keep safe inside himself long after he's left the museum. He sits back on his heels, nods to the statue—an odd gesture of, perhaps, respect—and moves on without me, winding through the crowds. I stand alone a while longer and stare at the goose bumps raised and rippled, carved by a tool onto Proserpine.

In other depictions of this myth, artists paint a weaker heroine. Dürer etches Proserpine (Proserpina to Bernini, Persephone to the Greeks) as a dizzying pinwheel of limbs, the center point of which are her breasts, bulging comically like bugged-out eyes. Alessandro Allori shows her placid and blank, seemingly bored by her kidnapping. Rubens bends her back over the edge of Pluto's speeding chariot, her will lost in the blur of momentum. Rembrandt's Proserpine limply claws at Pluto's face from a vacant state. Theodoor van Thulden leaves her stunned, head tilted up, arms skyward, as if asking for a better god to intervene and save her from her fate.

But Bernini's Proserpine is alive.

Her body is strong, and she torques it forcefully against the god, trying to free herself. She smashes the hardest part of her palm into Pluto's face. He grimaces. Bernini leaves Pluto dazed, off-balance, faltering, reminding us that Cupid's arrow kidnaps his agency, too. Ovid's myth tells of two forced transformations and Bernini shows us two people in motion, struggling unsuccessfully against their fate. The statue is bright, the brightest thing in the room, and it hums with the energy of the aggrieved—Pluto hurts Venus who hurts Pluto who hurts Proserpine; this circular hurt, placed on Proserpine's thigh, her stone flesh yielding below the god's grasp. It is stupefying: I am dimmed by awe, aversion, desire.

I've been standing too long and my right hip begins its familiar twinge. If I don't find a place to lie down, stretch, and rest, my body will start to lock up. The straps of my backpack are slightly uneven, and I can already feel the pressure causing the muscles on the right side of my curved spine to cramp.

I find my neutral room and count 1 2 3 *4* 5 6 7 *8*.

Slowly, I pass by the sculpture. There is other art to see here. I stop for balance and to rest. My back is stiff and stiffens more. Pain transforms the floor's stubborn slant; reorders it, distorts, unmoors; the plane is changed, both breadth and pitch. The pull of all this art is gone. Everything is just a thing now. Pain breaks my bond with all but it.

I look for an exit sign and walk into a new room and the stranger is there.

He faces away from me, but the line of his shoulder straightens my way, aims. He knows I'm near. I stare a moment too long, and he turns toward me. I flinch, recover, then move to study without attention the nearest painting, anything other than him. He watches me through the room. What a feeling this gives me. It electrifies the experience of looking elsewhere. I track his graceful maneuvering through the crowd. His long consideration of a piece of art, his eyes flicking up to a gilded ceiling—it is all for my benefit. I try to find enough energy to imagine a divergent reality, one in which I become a beautiful body blushing with desire, pain numbed, mind blank, dragged into the present moment by lust and left there, confused and alert. I follow him. He crosses the room, so I do, too; he turns a corner, I turn; he is steps ahead, his scent seeks me, the length of his neck is the length of my name; I am possessed, not by him, but by bloated, ornate reverie, by possibility. A curtain lifts to reveal a new narrative: a lady meets a stranger and now a real story can begin.

When I was six, I held my father's hand as he followed a red-haired woman around a department store. She was a stranger, but regarded my father with a knowledge I didn't understand. She looked at him until he lowered his eyes. She moved through the aisles, knowing he would follow, and he did. I walked behind my father, hidden from the woman, but she was not hidden from me. She wore a white dress, the precise image of which I can recall as if she were in front of me now in the Galleria Borghese. Delicate, loose, translucent. Often, I've fought the urge to buy something similar, wondering what effect a dress like that might have on me. My father had squeezed my hand. He'd whispered, *Keep up, keep up.*

I watch my stranger in the Galleria. Would I follow him out of the museum and into an imaginary night? He's ahead of me in the grand hall, *keep up, keep up*, but I can't keep up. My hip stops me. I rest against a wall. The pursuit is over, and I am, again, only myself: a tired mom, overheated and unable, unwilling, to keep walking. The stranger pauses in a far-off doorway, maybe waiting for me, but it's too late, my fantasy deflates, I'm beat, so beat, museums are exhausting, the day is done; the opening through which the unexpected could emerge is now closed, and I want to go home, or at least to the hotel and its air-conditioning.

I stand in the vestibule, just ahead of the exit, to get my fill of free Wi-Fi before leaving. I cycle through my email and social media accounts. A text pops up.

Isn't it a bit strange (it's my mother) *to go to Italy without telling anyone?*

I don't tell her the truth because I don't know it. Whatever it is will be embarrassing. If I tell her about Colin and Jay at the bar, what they'd said to me and what I'd said to them and how it had shifted something in me, how it had taken me from my family and put me on a plane to Rome—well, I can already see her rolling her eyes, heavy sighs all lined up and waiting.

Strange how? I text back.

My mother delivers disapproval in the form of questions. *What happened to your PhD?*

Nothing happened to it, I respond.

Should something have happened? Should work have happened?

Probably.

Dots undulate, bubble up, then dissolve into the depths below my cell phone screen. My mother quits the inquisition. I imagine her frowning at her screen, eyebrow raised. Her silence voices her real concerns: my son, my husband, my new job; the common thread: my abandoned priorities.

I put my phone in the pocket of my dress. I'm ready to leave. I grimace and bend at a drinking fountain. I feel the approach of a body behind me. A voice says, "Beautiful."

When I right myself, the stranger is there, staring.

"*Bellissimo? Bello?* You speak English?" the stranger continues. He's American. I nod and follow the trajectory of his hand as it rises and begins to gesture all around us. "Beautiful," he says.

"The museum?" I ask.

"Yes," he says. "Did you get in for free?" I try, a last hope, to twist this question into a pickup line, but it won't go.

My father tried one of his many pickup lines on the red-haired woman in the white dress in the department store. She was the physical opposite of my mother, not just because my mother's skin, eyes, hair were all darker, but also because of the way the red-haired woman moved. She floated as if underwater and she touched things just to touch. She fondled a mattress on display, used her hand to rub its edge, an errant caress to signal something, to put a scent into the air. My mother might have crisply flipped a sale tag or read a warranty, but she would not have engaged in this musky circling. The red-haired woman ran her fingers across the silky threads stitched into the mattress and then, under my father's careful gaze, she'd turned her hand over, exposing her soft, open palm. And my father stopped her. He held her at the elbow, fingers pressing in on soft, white skin.

The stranger is talking to me. He says, "This building itself, right?"

"Is what?" I ask. I've missed something.

"Is the most beautiful part, more beautiful than anything in it."

"I don't think so," I say.

"Do you think only art can be beautiful?" I hear a hiss and then a clicking sound, his tongue against his teeth. "Ah, ah, ah," he says, scolding me as dogs are scolded.

"No," I say.

"Only bodies?" He looks at me with discomfort for a moment.

Oh, I think, *I know what this is now.*

He's got that itch-in-the-brain look, like he's seen something go askew and he just needs to fix it, to fix something. My disability is obvious, but its details are unclear; to look at me is to feel information both shown and withheld. These ideas in opposition create cognitive dissonance and this makes people uncomfortable in a way not reducible to prejudice alone. There are patterns of reactions to this dissonance. People stare, mostly without realizing it. Some people cannot feel at ease around me until they know what they want to know. Once, at a restaurant, a woman snapped her fingers as I passed by her table and said, "Explain yourself."

The stranger needs to talk at me, needs to explain and label. He wants an uncertain thing to yield to a category assigned by his reason.

"What people don't realize is—" He keeps talking and I let him, but I've already retreated to the neutral room in my head where I'm having a different conversation with no one. He's performing a familiar soliloquy about how beauty standards are really made up by marketers and shift with the times and I'm nodding politely, waiting for the moment to pass. I've been here before and I know what comes next. In a minute he'll tell me beauty is in the eye of the beholder.

"But this building is objectively beautiful. Don't you agree?" he asks.

"No," I say. "I can't think of anything objectively beautiful."

"I think it's stunning. Don't you think it's stunning?"

"No, not this one."

"Well, it's all subjective, isn't it?"

"I don't think so, no."

"Isn't beauty in the eye of the beholder?"

"I don't think anyone who says this knows what it means."

"No?"

"Or rather, it has a meaning no one believes. It's a silencing sentence, one that reduces rather than explores one of the most exhilarating human experiences. The experience of beauty. What a shame."

Of course, my half of the dialogue happens only in my head. I am not a participant in the present moment. I do not want to talk to this guy. I nod and tell him, *Sure, the building is fine, it's great, it's beautiful*, and I wait in the neutral room for this conversation, a repeat of so many others, to end.

The stranger says, "My ticket cost a lot. You are lucky."

"Sorry?" I say.

"Don't tell me you paid? Didn't you see the sign when you entered? People with your situation get in free to most museums. Next time, I'll borrow a crutch." I smile and he smiles.

"Oh," he says, watching me. "I'm not being offensive. I work with people like you. I work *for* people like you." The stranger tells me his name is Joel. He's an acupuncturist from South Florida.

"I saw you in there before," he says, looking up and down the length of me again, "and I just wanted to ask—"

"Oh, no thank you—"

"What is your nomenclature?"

"My nomenclature?"

"Look," he says, "I'm not going for the three f's."

I want to know what the three f's are. I wait. He's hooked me and he knows it. He waits. We're at an impasse. His furry eyebrows, which seemed distinctive only moments ago, now look like fat caterpillars stuck in skin. Joel's black hair, soft as cat fur before, looks brittle and rough. He runs his hands through his hair and bits of skin flake and rain down on his shoulders. His fingernails, oily.

Joel says, "Fame, fortune, and favor."

"Ah," I say.

"You know what I mean?"

"I don't," I say.

"I'm not trying to sell you something. Not at all. I specialize in malfunction of extremities. I couldn't help but notice—"

"No, thank you." I turn my back and drink again from the water fountain.

"Sorry, it's a professional curiosity, if you could just—"

"No, thank you," I say.

"Excuse me," says Joel. He doesn't want to be interrupted. "I just wanted to help." He exhales through his nose. "I am offering to help you. I am actually willing to help you. You just have to tell me—"

People are quick to assure me that they are not intruders. They insist they are actually willing to help me. There are these oils, strangers tell me, these tinctures, herbs, powders, pills, yoga poses, meditation techniques, mantras, yodels, chants, supplements, hemp seeds, CBDs, drugs, gemstones, crystals, preachers, energy fixers, energy shifters, people who will realign all my energies, line them up just right; or, some will say, *Let me lay my hands upon you for I am a vessel of the Lord whose love will heal your body*, which is not the part I most want healed.

"I'm not a bad guy," says Joel from South Florida. He sees I am uncomfortable and that is making him uncomfortable, and he wants me to absolve him, and he wants me to help him believe, as everyone does, that all is pardoned by good intention.

"I understand," I say.

I want a dark room and a cool glass of water. Pain delays anger but it will find me later in the night. I let Joel hand me his business card. He presses it against my palm, leans in; a whisper on the wind, *If you're ever in Florida . . .* We part, the curtain falls. Later, I will replay this conversation over and over, thinking of better things to say, thinking of a better version of myself, a faster, smarter, more certain version. But in that moment in Rome, I was someone whose first and only impulse in the face of discomfort was to retreat, to leave my body nodding while the rest of me, the realest parts of me, waited in the neutral room. I did not see another choice because I was not yet aware I was making one.

I stand alone in the bathroom at the Galleria Borghese. I pat my cheeks with cold water. My face in the mirror: swollen, sweating, and red, and oh god, my awful face, flushed, foul, and confused. This face floated behind the stranger, followed him from room to room. Poor guy, poor Joel, tracked by a troll oozing a crimson, ghoulish lust. I am vibrating, but no, not me, my phone. My mother is texting again. I pull my phone from my pocket to read her text, but it dies in my hand. I am focused on all the wrong problems. For weeks I'd read up on art in Italy like there'd be an exam to pass later but had failed to imagine just physically being present in Rome and needing basic, sustaining things like dinner plans, a water bottle, a European plug adapter.

I'd not thought about practicality because I don't want practical things to happen. I want *an event*, something like a bar behind a phone booth, a séance at a village festival, a lever to pull, a secret revealed, a mountain guide to guide me, a mystic in the alley, a mystery to solve, a man on the run, anything, anything at all, but I am not of the ilk who discover such things. I am neither sensible nor adventurous. I am someone who can assemble an excellent bibliography and call it knowledge. I'd prepared for Rome by reading fat biographies of Bernini, accumulating piles of facts about the past, none of which would lead me to an experience in the present.

The afternoon sun pummels the façade of the Galleria. The tourists and I stand outside on the lawn and take pictures. A woman in a green hat, heavy camera in hand, breaks from her family, gestures to me, asks if I'd like her to take my picture in front of the museum. I shake my head no. I find a shady spot in the surrounding gardens and I lie down flat on my back and begin my stretches, crossing my right leg over my torso until a warm and relieving ache radiates through my hip. I stare at the sky. Something darkens the space above, the same woman in the hat leans over me and asks if I'm OK. I nod my head. Yes, I'm fine.

Sitting up, I see Joel standing in front of the Galleria Borghese, admiring it below a sun that bleaches them both. Perhaps Joel sees himself reflected. Two bodies, the Galleria and Joel, ivory and grand; two testaments to the enduring idea from the ancient Greeks and Romans that beauty is rooted in symmetry, measure, order. Perfect circles, straight lines, squares. Joel and the building, inscribed in the same circle of thought, no, the same repeating *circles* of thought, concentric, emanating out, over and over.

The Galleria's columns, its strict Palladian proportions, come from classical formalism, from the temples erected by the ancient Greeks, and from Vitruvius, whose *De architectura* is the only extant instructional architecture text from antiquity. Vitruvius tells us that man and building are best built in accordance with mathematical principles. Da Vinci's *Vitruvian Man* displays man's ideal proportions, perfectly inscribable within circle and square. The *Vitruvian Man* is a descendant of the *Doryphoros*, the masterpiece of Polykleitos, a beloved sculptor from fifth-century BCE Athens. Polykleitos's statue is of a spear bearer who is shown stepping forward, torso curved, his weight on the right leg, his left at ease, one hand curled around a phantom weapon.

In *The Canon*, a companion treatise, Polykleitos detailed the exact measurements of each part of the spear bearer's body as well as the precise distances between them. "Such perfection in proportion," wrote the physician Galen of the sculpture seven hundred years later, "comes about via an exact commensurability of all the body's parts to one another: of finger to finger and of these to the hand and wrist, of these to the forearm, of the forearm to the upper arm; of the equivalent parts of the leg; and to everything else."

Both the treatise and sculpture are lost, although there exists—in various stages of deterioration and imperfection—Roman copies remade in marble. The most famous copy was pulled from the ashes of Pompeii.

For the Greeks, these perfect proportions were not random but were drawn from an intricate and holy design observable in the natural world. So, orderliness in the human body was proof of that person's innate, divine harmony; to be beautiful was to have one's parts function together in perfect relation to a whole, just as parts of nature function together. Temple, torso, tree, leaf, wing, rose, all lined up by the eye of God; His patterns repeating everywhere: buildings built by the golden ratio; fractal branching in the trees above me; on the skin of the fruit that falls from the tree; on Joel's skin; and on the wings of the dragonfly that passes overhead: Voronoi tessellation.

Beauty could be caught and pinned by the regulating forces of design, measurement, order. Beauty could be whittled down to principles. Measure and proportion are everywhere identified with beauty and virtue; Plato wrote, "Beauty, proportion, and truth . . . considered as one."

But my eye gets bored traveling from one end of the Galleria to the other. Halfway through, I've seen all there is to see. Symmetry is predictable; I am soothed but not surprised. To say that beauty was merely the result of definite measurement deflated the mystery of the aesthetic experience: that bodily recognition, an ancient sense tuned to beauty, a physical seizing of beauty and of beauty's dissonance; a welcome fever, a palpitant thrill, pleasure ill at ease, a turned stomach, a chill, prickling hairs, goose bumps, high attention. And I have felt that high attention in the presence of art, people, ideas, sounds, storms, sentences, sunsets, streams and rivers and oceans, colors, efforts, failures, loss, pain, and how much of this can be measured? It is both there and not, neither subjective nor objective. I like the vastness. I want to keep the idea of beauty like a stone in my hand, turning it over and over.

But maybe I am dismissing the ancient ideals because they don't fit the story I tell myself about myself. My body did not fit into any narrative of order, proportion, plan. What was my lineage and where was it celebrated? In truth, I might find the Galleria building beautiful had I been born looking more like Joel or if I were at least touched, loved, fucked, chosen by this type of beauty. Maybe then I would submit to its rigid ideals if I were recognized as worthy of experiencing them.

Just as falsehoods threatened truth, disorder threatened beauty.

"Ugliness," warned Plutarch, "is immediately ready to come into being if only one chance element is omitted or inserted out of place."

The muscles around my spine throb and so I stay in the grass a while longer, willing the pain to subside. I lift the Bernini book, heavy as a brick, from my bag, knowing I can read for an amount of time determined solely by me. I can stay on my back for the exact number of minutes it takes for me to feel a bit better and I will not be embarrassed for how long it's taking and I am not delaying anyone because there is no one with me to delay, and if I'm triggering pity from passersby, I don't notice because I'm staring above at the sky again, free from the eyes of others, and I am so grateful to be alone.

People usually notice my height first. I'm short. Then they notice the way I walk, then that my legs from the knees down and my feet are underdeveloped and disproportionate to the rest of my body. My spine is curved, which makes my back arch forward. I have hip dysplasia, which means my hip joints are misaligned and unstable—the ball part of the joint grinds on a flat plane of bone in search of a socket that never formed. This hurts and I'm never not aware of it; pain plays a note I hear in all my waking moments. I walk by rolling my hips, which gives me a side-to-side gait. If I wear my hair in a long ponytail, it whips back and forth like a pendulum. I move slowly. I'm slow on stairs, but I can go up them if there is a railing. My arms are strong and I pull myself up as much as walk up stairs. The medical name for my disability is sacral agenesis. I was born without a sacrum, the bone that connects the spine to the pelvis. *Agenesis*, from the Greek, meaning a lack or failure to generate. My missing sacrum, my omitted element.

I want to explore more of the Borghese Gardens but the sun evaporates my plans, leaves me in a dazed state of generalized regret. Happy families pedal past me on canopied quadricycles. A man working the rental stand waves me over, then has second thoughts and nods an apology. I stand at a fountain, gulp water, pant, wonder how much of Rome I can skip. Joel and his eye on me had been a spike of excitement in an otherwise flatlined day.

I was supposed to be in Milan but had switched the ticket. I'd convinced myself it would be a failure beyond redemption to come all the way to Italy without seeing the Bernini sculptures in the Galleria Borghese. I held an operating belief that proximity to beauty was transformative, but what I might be transformed into—I'd not thought that part through. And now I have seen the Berninis and if I am changed it is not for the better. I remember the feeling of standing next to Joel, before he'd spoken to me. I remember looking with him at Proserpine.

I have a hunger to be elsewhere and otherwise. In an essay my father wrote right after my birth, he described himself as having a "motorcycle personality," meaning that he was unstable when idling but solid on the move. He wanted to be where things were in a state of becoming. He did not like to be firmly set in the present. He was fearful of stillness. I'd inherited this fear. I, too, want to go where everything is new, which means I always want to be somewhere other than where I am.

I walk toward the bus. People pass me on the sidewalk. I'm walking too slowly. I don't belong here. I don't know how to fit into the stream of forward movement. I feel the absurd weight of the Bernini biography in my backpack and am annoyed by myself and all my impulses.

I'd brought the book—and, to further the cliché, a notebook (a Moleskine!)—with the thought of lounging all day in the Borghese Gardens, reading and writing, a possibility that disintegrated in the first minute spent below the vile and reckless sun.

I'd been restless in the months following the night in the bar with Colin and Jay and I am restless still, but it strikes me as disingenuous and of a derivative variety, cribbed from the travel writers I'd read and admired who bore me no similarity. That literary genre offered mostly white, able-bodied, unencumbered men of means, many of whom were alchemists able to transform the busy strangers they met in foreign lands into drug-sharers, secret-havers, exotic sex partners. The few female travel writers permitted in print were usually mourning the death of something—their mothers, brothers, sisters, entire families, dogs, marriages. I have no transfiguring powers and nothing yet to mourn. I can only perform an embarrassing impersonation of someone else, someone suited to the experience of moving around the world. I cling to an uncritical certainty that I'm excluded from travel and its literature by body and bank account. But recently my circumstances had changed.

There is a kind of traveler who, upon returning home, languishes in the miseries of the trip. A fair number of this type end up in philosophy graduate programs. Others end up writers in New York. They can be sighted skulking the outer edges of readings or colloquiums and at the receptions that follow, where they'd frown and recount how their recent trip had been swallowed up by the (agonizing!) attempt to read a long (Russian) novel or a dense (Marxist) theoretical or historical (obscure, but of great importance) text, all while on the wrong train at the worst time.

They were also mostly white, able-bodied, unencumbered men of means, who lugged their Lukács from Brooklyn to Paris to Latvia to Bali to Budapest where they attended conferences or chipped away at dissertation research. They spent mornings moping at the archives and nights drinking at the bars. God, it was awful, they'd tell me with a wink. You wouldn't believe how terrible it was for them to be driven out of Spain by their lover's jealous boyfriend or groped through glory holes in Berlin or punched out at bachelor parties in Bratislava—utter agony!—but also all necessary! Necessary to the work of thinking, writing, of being serious, and, most importantly, necessary to the work of not becoming their fathers.

They'd return to New York when it was time to drink with New Yorkers and deliver lectures or readings to new audiences. These friends of mine, I guess, identical in their fawning joylessness, would tell the sensational morsels of their stories in flat tones, daring me into a wide-eyed reaction that would betray my innocence and lack of life experience. Now here I was doing my bad imitation of their bad imitation of the men we'd read about all our lives: the ascetics, the hermits, the flâneurs, the philosophers, the poets, the bards, the peripatetic Greeks who believed one had to be in motion to have deep thoughts; Wordsworth with all his walking; Socrates, Laurie Lee, Lawrence of Arabia; men who left normal life to go to the mountain, to sit on the hilltop, to roam the deserts, forests, glaciers, oceans; men who voyage alone, always alone, removed in mind, body, and spirit, kept clean of the mud of the world because only then could they finally, finally think clearly.

As a reaction to all I felt excluded from, I often found myself lifting my class status like a shield. My husband, Andrew, and I had both grown up in the rural Midwest. Andrew's parents had never married. He'd been raised by his mother, an itinerant pastor who moved yearly, shuffling Andrew from school to school, house to house, in one small, tornado-blown Missouri town after another. I'd been raised by my mother, a grade school teacher, on her farm in Kansas. Andrew and I moved to New York with four suitcases, three grand, and an infant. We had family support solely in the sense that our mothers loved us and did their best to stomach our choices. I started a philosophy PhD and adjuncted for pathetic wages at colleges across the tristate area, lesson planning and grading papers while on subways and commuter rails. My main access to the esteem of my peers—who were, with few exceptions, mostly young, white, able-bodied men with family money—was in how little I had despite how hard I worked. This, plus my disability, plus motherhood, bathed me in a tragic light, which, if I stepped into it just right, lit me up with the look of moral superiority. I could wield this as a weapon to ward off the axe swing of exclusion, and it would make me feel better for a moment but ultimately only deepened the cuts that kept me from others.

Andrew worked the kitchen in a sports bar in Manhattan, frying food for drunk fans who tipped if their teams won. His shifts ended in the early hours of morning. He slept on the subway back to Brooklyn, often missing his stop, waking with a kick from the conductor as the train stalled at Broadway Junction. He made it home in time for me to hand over Wolfgang and kiss him goodbye as I rushed out the door on my way to teach another class.

Being broke had a rhythm, a culture, codes, a language I spoke fluently, a narrative I understood. But then, a week before the night in the bar in Brooklyn with Jay and Colin, I'd been called into the dean's office at one of the schools where I adjuncted and was unceremoniously offered a full-time, salaried position. I'd completed a PhD in English before starting my second in philosophy, so I was qualified to cover a range of humanities courses and this made me a cheap hire for the school, which was small and just needed someone to teach a lot. The pay was fair, the benefits were fair, and, best of all, I didn't have to go out and sell myself on the academic job market. In the space of a minute, my most pressing problems disappeared. I felt light-headed, relieved, and then I felt a sadness that pinned me in my seat.

I tried to listen as the dean explained my new role, but her voice cut in and out and my vision winnowed. I felt relief, but also loss. I'd been working toward this exact goal for years and now my efforts had paid off, but instead of pride, I felt the nausea of inauthenticity. I'd entered the dean's office anxious about every dollar I spent and walked out middle class. I signed away a life I knew, one in which the trip I was on to Rome would have been inconceivable. This job gave me and my family a stability we so desperately needed but our shifted class status itched me like new skin healing over a wound. To enjoy the pleasures of money felt impossible, but to take this new security for granted was morally abhorrent, and I saw no easy place to land between these two feelings and so I bounced uncomfortably from one to the other.

The bus takes me away from the Borghese Gardens. I try to continue reading the interminable biography of Bernini. In truth, I like combing through it paragraph by opaque paragraph. Academic philosophy had trained me to read slowly and badly and with the belief that what I was really doing was grinding out a path to enlightenment, a belief that thrilled me. The whole project of philosophy was to get nowhere. If, by chance, a philosopher did discover the answer to a question—say, how to solve for the hypotenuse of any triangle—it was no longer philosophy. Facts got kicked out to the other disciplines. Philosophy, by its very nature, required uncertainty.

To seek the truth required one to endure dissonance, and the ability to sit with this discomfort was what separated the philosopher from other people, or so we're taught in graduate school. I bought into this image of the philosopher and his work and willed it to bleed into the rest of my thinking. I like the thought that what I seek will be discovered if only I can withstand what others cannot, that pain has purpose, that I'm not lost, but just on the harder path.

At a restaurant near my hotel, I eat *cacio e pepe* because it is Rome and obligatory to eat *cacio e pepe*. I eat it alone and think, *I am here, here I am, still me*. I bring the biography of Bernini and I try to read, but the words blur on the page. I put it down as someone comes to fill my water glass. The waiter asks me what I'm reading and I show the cover and he coos approval.

"God's favorite," he says. "Very important to Rome, to Catholics."

He's impressed I'm not taking pictures of the *cacio e pepe* because American tourists, he says, come to the restaurant and take so many pictures of the *cacio e pepe* that they eat it cold and complain. I do not say the truth: my phone is dead. I want to text my mom and I want to take a picture of my *cacio e pepe*. Instead, I read my book. The waiter nods approvingly. I don't get far. In my head, I keep repeating Bertrand Russell's line, "Rome had no new ideas," by which he meant that philosophy in the Roman period was not free, but was molded to fit an ideology, namely Christianity. Art followed. The famous works on view here are largely the result of papal commissions. The Galleria Borghese was the collection of a great patron of the arts, Cardinal Scipione Borghese. Art was for God and beauty was from God and God was all over Rome and Rome had no new ideas and neither did I. But then a new idea comes.

When the waiter returns, I ask if these American tourists ever left things behind. He returns with a cardboard box. I fish around until I find what I need.

"With our compliments," the waiter says. An adapter. I plug in, I am resourceful, I have solved a problem without making a reading list first. My phone charges, lights up, alive. I text my mom. I send her a quick pic of the end of my pasta.

I leave the restaurant after the sun has set. Rome is dark. I'm tired and need the shortest route to my hotel so I cut down a dim alley.

The road turns rough. I trip along the way. I keep my head down, eyes squinting at my path, and so I don't see the men first but hear them. They're laughing. I move to one side of the alley and they move to the same side. I step the other way and so do they. There are four of them. I hear one speaking to me, but I don't know what he is saying.

Their interest in me, their sound, turns me stony. I open my mouth and out comes not words, but strained guttural notes.

One man jogs past to stand behind me. Another puts his hand on my shoulder and backs me up, toward the wall, toward his friend. His friend is tall. They want to take my picture standing next to him. I'm short, a dwarf, which is funny, hysterical. I'm not real. Just a strange thing in the alley. The flash of their camera. I freeze. Then I'm back in the dark.

When I was a teenager, a man once watched me going up some stairs and he said, "Grace eludes you." I seemed to be struggling, which struck him, I suppose, as ugly.

Does this man remember what he said to me? Does he return to the memory each time he sees stairs?

I still—two decades after this man watched me walk up the stairs—step aside to tie my shoe to allow people to go ahead of me. I fake phone calls so that others will walk up without me. I pretend to wait for someone who isn't coming. I bide my time, clinging to my weak ruse of self-protection, until no one is looking. I do not climb stairs until I can do so unobserved. I've never stopped preparing for the next person who will see me walk and deny me grace.

The way words stay, the way sentences stay, the way memories invade my present, the way a stranger looks at me and speaks: shards that become a mirror.

In Rome, men block my path. They are drunk. The tall one wants to leave, done with this picture project. Another man drops his phone. His friends laugh at his clumsiness. One taps the other's chest and just like that they're distracted by a new plan, a diverting interest, and they leave me without further incident and carry on with their night, never to think of this moment again.

Midnight in Rome, dinnertime in Brooklyn. My family appears within my phone's frame with their bowls of half-eaten spaghetti. Wolfgang's mouth is red at the corners, sauce on his bare chest; his little face blurs, turns to blocks, stops, then refocuses itself on the screen. My husband tells me about his day and what he'll do the next and what Wolfgang had done that day and their voices are sweet and clear and I am grateful and numb.

"How's Milan?" Andrew asks.

"I'm in Rome, actually," I say.

"You are?"

"Yeah. I had to see the Berninis."

"Sure," he says. "Makes sense."

I don't tell him about the men on the street. I've learned not to share these experiences, especially with able-bodied people who can be quick to tell me how I should feel, that I should *just ignore it* or *learn to laugh it off* or that I'm *being too sensitive* and *it's not a big deal* or *it is a big deal*, *a huge deal*, and I should be angrier, much angrier, *why aren't you angrier?* Often these statements are made with good intentions, meant to embolden me or shuttle me through the encounter, but they always have the opposite effect, leaving me feeling chastened and misunderstood. It is a deft act of erasure to be told how to process a situation by a person who would never experience it.

This is not why I withhold the scene in the alley from Andrew, whose capacity for empathy is wide and manifests mostly in the form of careful listening. He does not pretend to know what he cannot know and he never tells me who to be or how to feel.

I don't tell Andrew about the incident because I don't want him to feel I need help he can't give me. I don't want him to worry or be afraid. But more than this, I don't want my hurt to spill onto him, staining him. I want to protect him from that part of me, and Wolfgang, too, and so I keep it to myself, but in doing so keep my distance, failing to share my life with them. Wolfgang starts to talk, but there's a mechanical screeching coming from the screen. I turn down the volume on my phone. They are both very far away.

I find Ovid online and reread Proserpine's story. I'd forgotten Cyane, the naiad, who rises from the river to beg Pluto not to take Proserpine. He ignores her, and she feels so much sorrow, watching them descend into the mouth of hell, that her body disintegrates.

In silence carried in her heart a wound beyond consoling. . . . You might have seen her limbs soften, her bones begin to bend, her nails losing their hardness. . . . Shoulders, back, and breast dissolved, disappeared; and, in place of warm and living blood, water flows.

I recognize in Cyane's fate an appealing and familiar way to solve a problem. I draw a bath and believe, too, that all my sorrows will be resolved by disembodiment. My legs and hips throb. I slip, stiff as marble, into the hot water. I welcome the release, the unmingling of mind from body.

In the department store, my father had approached the red-haired woman and he'd held her by the arm, his fingers pressing gently into her skin. We'd followed her from room to room, past stacks of shoe boxes, racks of clothing. My father pretended to look at a blender. The red-haired woman showed no interest in buying anything. She'd looked at him, eyes bright, smiling. He touched her elbow.

My father might not remember this moment, this flirtation with the red-haired woman. She was not the woman he eventually left my mother for or even one of the women he cheated on her with. This was just some brief encounter on a Tuesday or Wednesday, a lazy afternoon in Kansas, in summer. We'd gone to that department store on an errand for my mother. I don't remember what she'd needed, a garden hose or a gallon of paint, some ordinary object she'd use in her constant effort to keep up our home.

My mother asked of everything: What work needed doing?

My father was after grander experiences. He craved excitement, surprise. He wanted the beauty that sharpened the edges of heightened feeling. None of this was easy to access at a department store, grocery store, pharmacy, doctor's office. He failed to find value in life's bland tasks and resented the fact that such work was expected of him.

It has been ten years since I've seen my father. Three months before this night in Rome, I received a letter from him. He was getting sober, it read, and he had entered a seminary that gave him the chance to sit in solitary reflection. He'd mostly been thinking about happiness, a concept that remained elusive to him. *I was happy once*, he wrote, *or maybe I am happy now. I don't know.* He found himself, more and more, retreating to scenes from our shared past: us in our truck, kicking up dust on the backcountry roads that led to our farm, him behind the wheel, a child-me next to him in the passenger seat. It was there, in the memory of the quotidian drive home, that he located a feeling like happiness. *Sometimes our dog Angus, that wonderfully iridescent black Lab mutt, is with us*, my father wrote, *sitting in the bed of the old, beat-up Ford 350 we used for carrying hay and pulling our horses. We roll the windows down and let in the dirt from the road until we choke.*

My happiest memory came when Wolfgang was four months old. On the day of his birth, I felt pain, anxiety, exhaustion, terror. Joy was noticeably absent. I waited for it in the weeks that followed but a depression came in its place. I hoped it might at least blunt the dread that kept me awake all night, listening for sounds of my son in his crib, but instead my fear was made brighter by it, just as stars shine against the black sky and by contrast are brightened.

Months passed and then, one day, postpartum darkness gave way to a silvery dusk and, on one of these lighter mornings, I lifted Wolfgang from his crib and cradled him in my arms and made a silly face and he laughed—he really laughed. He looked right at me and emitted this high-pitched screechy music, this pure baby glee, and I heard the sound of it and I felt it, too; it spread throughout my chest, bringing with it a joy I'd never known before, a pure distillation of feeling that I can only describe as being shot through by some sacred beam. Now illuminated: this laughter, this piercing shriek and the one-toothed smile that propelled it out of him and into me, was the most beauty I'd ever experienced, and I was happy, but then he stopped laughing and cried to be nursed and so I nursed him and then I changed his diaper and, later, I laundered his soiled sheets and I dressed him and took him to the bodega and to the pharmacy and the playground and eventually I made lunch and dinner and, throughout all this, I was acutely aware of how quickly the experience of beauty dissipates and is replaced by boredom and the dullness of obligation.

My father's letter to me was an attempt to explain why he could not keep hold of the happiness he felt when we were together decades ago. He was cursed, he wrote, and believed I'd inherited this curse. He saw us as having a dual nature: one side shaped by prejudice, impoverishment, and self-delusion; the other side capable of creation, joy, song, sublimity. This duality made us sensitive and uncompromising, he explained. *I was happy and incredibly miserable at the same time.* We saw our authentic selves residing only in the romanticized abstract realms of art, creation, beauty—an interior space of thought and private feeling. We were petulant and impatient, claiming to experience outsized agony, when expected to endure less romantic realities. *We are people*, he wrote, *who see the world differently, who feel oppressed by the dreadful normalcy of life, who long for something more, something more beautiful.*

My father worked a government job that involved a lot of paperwork and no art. It put him face-to-face daily with Dreadful Normalcy itself. The fact that we needed money and that he was beholden to this work offended his deepest sensibilities. *Where do you go*, he asked in his letter, *to find escape from a reality that is oppressive to the soul?*

Now, in a bathtub in Rome, after the long day at the Galleria, I think of my father's letter. I see my pursuit of Joel as a way out of reality. Pain oppressed me, held me at a remove, filtered my every experience through the lens of itself, turned a vacation into an endurance activity, made it impossible for me to access what I really longed for, which was to be fully present in nearness to beauty.

Where do you go to escape reality? I went to alcohol and the random affair, my father wrote in his letter. It was an admission that underplayed the pain those escapes cost him and me and my mother and others, but it also obscured the truer answer to his question. He escaped most deeply into art, imagination. He turned tasks, all our banal outings, into adventures. He made up games for us to play in the aisles of grocery stores. He flirted with strangers, constructed grand romantic relationships from a few side-glances. Through his eyes the mundane became mythical. And I loved it.

I idolized my father and mistook his reluctance to face life's hard facts as a kind of nobility. But also, he was just more fun, certainly more fun than my chore-oriented mother. It was fun to run errands with my father, who would abandon the errand altogether if its realities started to infringe too much on the fantasy he'd needed it to be. And I am the same. Me and my father, looking to escape. Me and Cyane, dissolving in water, free from the need for further complex thought or action. I meet reality by trying to either transcend it or sit below its surface. I want the fix, but not the work. I want the world, but not its facts. I do not know how to reconcile opposing desires, to hold them in my mind at the same time. The attempt brings only weakness and I feel myself succumbing to the dismissal of the dissonant mind.

The curse, he explained, *our shared curse*, was this: we could not gracefully integrate our authentic selves with the stark facts of reality, and the inability to resolve this tension causes us to suffer and to make other people suffer. *It is not an oversimplification*, my father wrote, *to say that this tension between the "romanticized" and "real" world is the story of our lives.*

At the end of his letter, with a sincerity I could feel across the cold distance of prose on a page, he asked me, *How can I live an authentic life in the present?*

My father could not do this and feared I could not either, and I know he is right, we are the same, we share the curse, but I have one tool he didn't have: his example. His choices, and where they led him. I know my father's fate.

Proserpine is rescued in part. While in the Underworld, Proserpine ate pomegranate seeds, which means she has to stay with Pluto for part of the year and can return above for the other part of the year. The myth gives an explanation for the seasons. Winter is when she is below, and the earth mourns her absence by turning cold and hard; when she returns, the world warms. Proserpine is glad when she hears of this split.

Straightaway her heart and features are transformed / That face which even Pluto must have found / Unhappy beams with joy, as when the sun / long lost and hidden in the clouds and rain / Rides forth in triumph from the clouds again.

She wants to live half in darkness, half in light.

It will be years before I understand the ending and when I do, I'll see something more in the Bernini, too. It's not Bernini's technical skill that makes his sculpture beautiful. It's not its proportionality. It's that he keeps its tension apparent. Bernini leaves the scene uneasy, its dueling desires visible.

Proserpine, by the end of the story, is transformed. She belongs to two worlds, rules two empires: queen in hell, as well as above. Proserpine, from *proserpere*: to emerge, to slowly go forward. Bernini's fingers press upon her soft skin, building disquiet below the marble's surface, leaving there for you to see: a woman divided, a woman changing, yielding beneath the wrath of love.

2

A Dog in the Yard

A cry in the night, Kathmandu, 1983.

My mother heard it first.

She sat up and listened, eyes closed, and she tried to swim out from the confusion of sleep to land on what, precisely, she was hearing.

A single howl multiplied, echoed off the surrounding mountains. She woke my father and they listened in the dark and perhaps I heard it, too, as I was there between them, safe in my mother's stomach, a month away from being born.

The sound was a dog in the yard, minutes or hours away from dying. It lay on its side, eyes glassy with fear. The dog's stomach had been torn open by something, a phantom claw. In pain it made an absolute sound, a keening song. My mother and father stood in the yard together, stilled by the certainty of what they were seeing: a whole and seamless suffering. Both wished a quick death for the dog, but neither could deliver it.

When I imagine this moment, I see my parents on opposite sides of the lawn, processing the problem of the dying dog separately. I'm likely only half-right. They probably stood side by side but looked straight ahead and not to each other for an answer. My parents never seemed truly *together.* There are dates written on legal documents that mark a start and an end to their marriage and there were years we lived under the same roof and others when my father, in the midst of one affair or another, lived apart from us. But these distinctions don't hold much weight in my memory. They were always two planets whose orbits moved them sometimes closer together, sometimes further apart. But maybe it had been otherwise between them before me, before the dog in the yard.

My parents were similar in one crucial regard: they both considered with great moral seriousness the question of what it meant to live a good life. Their answers evolved in direct opposition and this might have been precisely what attracted them to the other. They could learn a new way to do the right thing. My mother, always practical, started thinking of what tools—a bat, a frying pan, a kitchen knife—they had that might help to hasten this dog to its fate, and my father looked up at the stars and started philosophizing. Was he justified in delivering a death that ended suffering? Could killing be an act of grace?

Kathmandu was overrun with homeless animals, thin from starvation, mangled in some way, an eye or ear missing. The longer these lonely animals lived, the more likely they were to make more lonely animals, and so the Nepali government would send out sweeps to collect the strays and euthanize them, and the streets would seem empty for a time, but they always came back. It was something you learned to accept, my mother had told me. If people are poor, animals will be poorer.

My parents fed and cared for many of the animals who wandered into their yard, wounded and hungry, but there was no saving this dog. My mother said they might crush its head, but they didn't own a car to run it over. She couldn't stand to see the dog suffering on and on. She wished for a gun; my father would have preferred a bolt of lightning. Finally, he said he'd drown it.

My father was always singing. He appears in my mind mid-solo. He sang Bruce Springsteen songs. He sang Elvis Costello, David Byrne, Otis Redding. He sang Richard Thompson songs so often and with such devotion that I can only think of Richard Thompson, beloved British songwriter, as a sort of imaginary friend of my father's. He had a singular receptiveness to aesthetic possibility. He could siphon it from the air. Every surface a drum set, every cylinder a microphone, every place a place to dance. We liked to watch TV together, but at commercial breaks my father would hit mute, turn to me, and recite a poem. He wrote bits and pieces of his own songs, his own stanzas, parts of short stories, chapters of books. He was always in the early stages of dreaming something up. To be immersed in creative action, to be constantly making something beautiful—this was what it meant to live a good life. Destruction of life was the end of beauty and the unmaking of things threatened all that was good. So, to drown the dying dog must have been, for him, an act layered with many horrors.

I can see the Tiber River from my hotel window in Rome. Peaks of meager waves ease upward, reflect a little streetlight, then dissolve back into the black. I see my father drawing the bath for the wounded dog in Kathmandu, carrying it there. My father, so tenderhearted, so sensitive to life—his same receptors that were open to beauty were as open to suffering.

My father had asked my mother to stay with him at the edge of the tub in their bathroom in Kathmandu. She was, after all, the one who dealt with life's hard facts, so she would handle this one. He needed this from her. He liked being at the edge of life's emotional cliffs, always curious about what lies in the depths below, but he didn't want the fall. He needed to stay secured to a safe place and my mother was his tether.

But she surprised him. She couldn't stay and help him kill the dog. She hid in the bedroom, covered her eyes and ears. And so, he pushed the dog's head under the water himself and as he did, a latch that held him to my mother became unfastened. She'd betrayed him by making him switch roles with her. He was never able to forgive her for that. He couldn't forgive being made to bear this burden alone.

My mother was also susceptible to suffering but she possessed the discipline to numb herself. *I tend to tamp down my feelings*, she's said to me many times, always with an equal measure of resignation and pride. Except she couldn't sublimate herself enough to kill this dog in Kathmandu. Maybe this was my fault. She was focused entirely on me and my unknown but certainly abnormal condition. Other mothers in Kathmandu told mine stories about their babies kicking and flipping in their bellies, keeping them awake at night, but I was due any day and my mother had not felt me so much as shift in her womb. I was an ominous mystery, heavy and lifeless. My father, who had no greater fear than stillness, did not know I hadn't moved. My mother did not tell him. She knew I was not real to him yet—I was only a concept, a swelling under the skin. She chose to give me all her strength, leaving none for my father that night; her orbit carried her closer to me and further from him in the months, weeks, days leading up to my birth. I was real to my mother and she was a woman who could face hard facts, and perhaps the possibility that I'd require more sacrifice, more chores, than she'd anticipated made me even more real.

I open my hotel window and in come mosquitos, a viscous wind, the brine of the Tiber. Across the river stands the Basilica di Santa Maria in Cosmedin; the yawning arm of its bell tower stretches into the soft night sky. Beyond the Basilica is Palatine Hill, and beyond it is the glow of the Colosseum, which stayed open through the night for smart tourists who did their sightseeing after the glaring eye of the July sun was shut. What did my father look at when he drowned the dog? I never asked him. Did he stare at the grout between the tiles lining the tub or did he look only at the water, his clear and quiet instrument? When he drew the bath, did he adjust the temperature until it felt neither hot nor cold, but like nothing at all? And when he held the dog down until it stopped moving against his palms, what had my father felt? I admired his great sensitivity and I envied it, but I, too, wanted only to walk to its edge and not go over it. I didn't want the attendant despair. I was inconsistent this way, unfair. I worshipped him, but I didn't want to know him truly and I didn't accept him as he was. I imagine him leaning over the tub, lonely and scared, watching blood blossom like water roses as the dog's wounds washed clean. Did he, my father, a man who found life's meaning in aesthetic experiences—did he try to find beauty there?

There were no hospitals in Kathmandu and so my parents decamped to Bangkok. It was June, hot. There was a movie theater that played American and European films from the seventies. They'd seen all these movies already, and they were dubbed in Thai, but the theater had air-conditioning. My mother's water broke in the middle of *The Godfather: Part II*. She rode backward on my father's motorcycle to the hospital, her pregnant belly exposed to the night air rushing past. There were dogs and cats roaming the halls. I was in breech position and wasn't moving and the umbilical—

"Not true," my mother says, cutting me off. "Where do you get these stories?" She snorts into the phone. Her breath crackles through the speaker pressed tight against my ear. I'd been so lonely for her, looking out my hotel window.

It's evening in Kansas, early morning in Rome. I hear a metallic melody, the sound of water being wrung from a rag over the washbasin. My mother is cleaning the dinner dishes. I close my eyes and I can see my mother clearly, standing alone in the kitchen, moonlight making bright the rims of wet plates drying on the rack, the farmhouse around her lit by the one lamp on in the living room. Her farmhouse stands in the center of a grassy clearing and beyond the clearing, enclosing it in a tight circle, are the crumpled litter of trees; and the Tiber River outside my window, an oil spill beneath my mother's same gleaming moon, becomes a black canvas upon which my mind's eye paints the image of—with all its familiar chips and dings—the white porcelain washbasin in my mother's kitchen and in the basin, her mugs and glasses, her hands.

"Which part isn't true?" I ask.

"There was no motorcycle. In Nepal, yes, he had a motorcycle, but he didn't bring it to Bangkok. He told you that?"

"He put you backward on the motorcycle—"

"A nice romantic image, but false," she says.

It was a nice image, one I felt attached to, the vision of my pregnant mother, beautiful and free, floating on the back of the motorcycle, watching the road grow longer and longer—

"What about the taxi driver?" my mother cuts in.

"What taxi driver?"

"You don't know this story?"

"Mom," I say, "if it's your story, I don't know it."

"Hmm," she says.

"Mom?"

"What?"

"Taxi?"

"My water broke in the shower at our hotel. Your dad hailed a taxi to take us to the Bangkok Nursing Home and as soon as the driver saw me, he tripled the price."

"Did you pay it?"

"No, I said, *Screw you.* This guy wanted to haggle, but I wasn't going to haggle on the day of my daughter's birth."

"What did you do?"

"I walked."

"How—"

"Slowly."

"No, how far, Mom?"

"Some distance. I remember grabbing a fence. There was no motorcycle. I walked! You don't know this part? How I waddled my way to the hospital?"

My father could read people well. He understood expectations. He saw what people felt they were owed and what they really wanted and could alternate, giving one or the other, but never both. This made him a brilliant storyteller, a dream guest at a dinner party. The seat next to him was a place of honor. He made people laugh and he made people believe that he was the brightest thing in any given space and approaching him drew them nearer a true center. I felt this way, too. As a child, I never wanted to leave his side. I was his most devoted audience member and was rewarded with his stories: how he got lost in Asia's labyrinthine bazaars, joined a motorcycle gang, played in a band. He told me about the Sherpas who led him up Everest and how thin the air had felt in his lungs, how it had focused him on nothing beyond surviving the mountain.

He had been the smartest person in his family, the smartest person he knew, and then he'd gone on to be smart in college and grad school and this certainty of his intellect made him feel like life's debts were paid and all he had left to do was stroll into his dazzling future.

My mother's physical beauty was her entry into my father's stories. She was, he'd told me many times, the most beautiful woman in the world. He'd seen her first across a conference hall. She was young, long dark hair, dark eyes. She was being recruited to teach in areas in the United States that needed teachers, and he was the recruiter. He walked across the hall to her and, with no introduction, asked her to meet him later that evening. She agreed and he left.

When the day's recruitment sessions were through, she'd paused in the lobby where he'd asked her to wait for him, but he was not there. She gave him a minute more and then another minute but still he did not come. She stayed in the lobby until the lights went out and then she went back to her hotel.

"Why did you wait?" I ask her.

She exhales into the phone to let me know I'm asking an unpleasant question, not for its subject matter, but simply because it is about a moment that has already passed. She's getting restless. Her voice cuts in and out as she, to keep her hands free, grips the receiver between shoulder and chin, muffling it against her cheek. She'd like to end the call but is short a reason since she knows that I know her chore order. She has already trudged across the field in her black barn boots to muck the stalls and lay hay flakes in the feedbags. Her horses, Echo and Jimmy, have by now been brushed and scratched below their fly masks, tucked in till morning. I listen as she moves around the kitchen, looking for more work to do. Her own dinner and the cleaning of dishes comes last, and now the dishes are done. But there's always another chore to find, and my mother finds them all.

"Mom, why did you wait for him?" I ask again.

"I don't know why. Why not? Who cares?" she says.

She sat on her hotel bed waiting and then there was a knock. My father, a stranger, was leaning in her doorframe, a raised eyebrow, a face made stony by pride, and he smiled as he explained that he was sorry to have missed her in the lobby, he'd been detained at a meeting with other recruits and, as happens sometimes, the meeting had evolved into an orgy.

"Were you mad?" I ask.

"I might not have known his name at that point," my mom tells me.

He stood in her doorway and told her about the orgy in a flat tone, daring her into a wide-eyed reaction, mistaking her for someone who might be impressed. She regarded him with detachment, clocked his posturing. She was and is a person perfectly alone, sealed up and solitary, taking up exactly the space she was due and no more. She watched his blue eyes shift. She unnerved my father, turned him sheepish. He wondered what her beauty might mean for him, how it might transform him into a person of visible worth, made shiny in proximity. Could she forgive him? Sure, why not, who cares, she could.

"Why are we talking about all this now?" asks my mom.

"I'm recording this conversation, by the way," I say.

"You better not be."

"Just tell me why, after all that, would you accept the date with Dad?"

"Hmm." She utters this sound angrily and then makes it again, *hmm*, but this time the utterance sounds like a shrug. We are quiet for a while.

Finally, she says, "You're supposed to tell people you're recording them before you record them."

"I know."

She's wary of me. She's bored. Revisiting the past wastes her energy. She prefers her chores, all of which are in the present. To get her to talk, I need a prompt that is the equivalent of knocking a picture frame lopsided. She'll feel compelled to level it.

"I only know Dad's versions of stories," I say.

The air coming in through my open hotel window is cool. The dust has ceased its swirling. The cicadas' pulsing drone grows louder. I listen to my mother rinsing something in the sink. For a moment, I'm not in Rome, but with her. I am there in Kansas beside her. I see what she's seeing and I see the frown on her face, I hear her run the tap, then shake out the rag again; I know exactly the rag, she's wiping down her kitchen counters, I know which side she'll start cleaning first and where she'll stop, I'm with her, she murmurs.

"Hmm, OK," says my mother. "Go on. Better let me fact-check."

My father had been afraid at first to go to Asia. My mother's side of the family was from the Philippines but had immigrated to the States before she was born, and she'd grown up in Kansas. She and her sister, Georgeanne, wanted to travel. They'd both found teaching jobs in an international school in Kathmandu. When my mother announced she was moving to Nepal, my father told her she was nuts, and she'd said he could come or not, but either way, she was going and unlikely to come back anytime soon. Marrying simplified their travel logistics, so my mother borrowed a skirt from her sister and my parents wed swiftly with few in attendance.

Nepal's lax drug laws made it a popular place for people who wanted to escape something. My father always had a crowd with whom he could burrow into the night. He liked drugs, liked women, but mostly he liked to drink. He was just barely thirty. He could easily envision a life forever overseas, roaming from one country to another, never stopping for long, never settling. His favorite movie was *Lawrence of Arabia*. In his heart, he was T. E. Lawrence, a brave stranger in a strange land, a complicated leader, a man of divided allegiances. He was after an adventure of the highest order and finally he'd found it, but it wasn't what he'd envisioned. He'd missed the apex, the true height of the hippie utopia abroad. He was a man who liked to be in the midst of a grand narrative but was somehow always a step outside of it. My father's ambulatory unhappiness circled my mother. Whose fault could it be but hers when things went wrong? She and no one else had dragged him out to the deteriorated edges of this massive continent.

I believe my father agreed to go to Asia because he loved my mother, but also because he didn't want her to live a fuller life than the one he would live. He wanted to stick it to his youth, too. He'd grown up poor and unpopular in rural Pennsylvania. He'd watched his childhood bullies become their fathers, going to work with them in the town's coal mines. He pitied them from Nepal. While drug addiction and alcoholism decimated the remainder of his Pennsylvania family, he was certain his own alcoholism had a noble quality. He could become an adventurer with vices that, against the backdrop of the Himalayas, could be recognized as fashionable, literary, and, most importantly, he would not become his father, who was dead, dead of a heart attack the month before he'd been born.

All he knew of his dead father was that he had been "a saint" and that his life would have been substantially better had this saint stayed alive to guide him. My grandmother never spoke of her dead husband, finding it either too painful or impractical or, most likely, both. My grandmother never remarried, never even dated. She clung to the ghost of my grandfather but did not keep his memory alive. She did not tell my father who his had been, what he'd liked, what he'd believed, how he'd treated her, how he'd cared for his family. My father felt he'd been robbed of his most essential model of how to be.

My father had one sister, Sarah. She was twenty years his senior. They never lived under the same roof and they were not close. She'd had a whole childhood with my grandfather and my father hadn't been given a single day with him. My father grew up jealous of women, the two closest in his life having had a lifetime of guidance from my grandfather, the vanished saint. This loss was a deep hurt, and he never let go of it.

My aunt Sarah visited us once in Kansas after we'd moved there from Nepal. I don't know what, but something in Sarah's childhood set her up for an adulthood filled with violent and hateful men. Her first husband tried to kill her. Her second was a wealthy widower who despised her and fought with her bitterly in public. This was one of my father's most visceral memories from his childhood, watching Sarah's second husband scream at her in restaurants and grocery stores. For the rest of his life, he had a fear of any form of public discord.

When Sarah came to stay with us, she spent most of her visit in a state of hypervigilance, anticipating danger everywhere. She worried so much about the possibility of me slipping in the shower that she lined its slick porcelain floor with all the bath towels we had. She insisted on holding my hand whenever we were in public, squeezing it until she left bruises. She called the police after observing a young black man go across our lawn, up our driveway, and into a shed in our backyard. Later, when the cops arrived, my father stood between them and our young neighbor, whom he paid to mow our yard, furious and cursing at Aunt Sarah, apologizing to everyone else. He was humiliated by this incident unfolding out in the open on his lawn and he yelled at Aunt Sarah, returning the hurt. She wept, not understanding how she'd become the villain in the scene.

My father believed, along with the Greek philosophers he revered, that ignorance was, allegorically, a cave, one that we could be freed from if only someone would break our chains and walk us from darkness and into the light. But my father felt he'd had no guide. He'd been given no god, no father, no teacher—only a grieving mother and distant sister, both of whom seemed ruined by marriage and by their refusal to leave West Pittston, Pennsylvania—he called it *the wasteland of their youth*—which he fled physically at seventeen. Mentally, he was less successful.

He wanted a family of his own, he wanted us, my mother and me, but he could not leave his childhood behind—it lived in his present, disrupted his future. He did not have a model for familial coherence, only dissolution. He could only reenact what was familiar to him, which was absence. He tried to change, I really think he did. He read obsessively, searching in books for the guidance he'd been denied in life. We were the same this way. We both felt we had no compass and so we clung to whatever sense of direction we could glean from books, art, ideas, from the tales of brave and brilliant men. My father searched in books for his lost father and I looked there, too, for mine, and we were both always alone, seeking the ghost.

I was born a ball of twisted muscle and tucked bone. My clubbed and lifeless feet were stuck up at my ears. I was bent in half. My legs were shriveled and short. The balls of my hip bones protruded, unglued from their sockets. My mother was unconscious, under anesthesia from the emergency C-section, and so my father was left alone with the sight of me.

My father wanted to travel to every country and speak every language. He was certain he had great art in him, certain he had ten novels to write. He'd made room in his narrative for a child who traveled with him, witnessing and recording these adventures, who'd tell tales long after of the great and complex father. There was no place in the arc of his imagined life for a body like mine. So, whose child was this?

My mother looked for me when she woke from the anesthesia, but I wasn't there. She said to my father, "Where is my baby?" He said nothing. I've imagined this moment between my parents: my father looking down at my mother, begrudging her those last seconds in a pure and liminal space where she knew I existed, but did not know the manner of my existence. I imagined the safe space dissolving when the nurses handed me over, mangled, red, and raw, yowling with need.

The doctors in Bangkok were stumped. Sacral agenesis is rare and they'd not seen it before and didn't know how to explain it to my parents. They could not name what I was, but they recited a list of what they guessed I'd never do. They said I'd never walk, never stand unsupported, never have a pain-free life, if I lived. They said I'd not be able to eliminate on my own—that's the way the doctors put it to my parents. My father pictured a life of emptying my waste. They said a disarticulation, the surgical removal of my lower limbs at the hip, was likely. This list shifted, shrinking and expanding over the years. Some things dropped off the list when proven false but were replaced by others. They told my mother I'd never be able to get pregnant. There were so few cases of sacral agenesis in existence, so few models, nothing to research. My parents listened to the doctors, believed their predictions, and later I did, too.

3

Go, Thoughts, on Golden Wings

My bus stutters down Via Merulana, a lush avenue vibrant with green and sprouting earth on either side. People shop at fruit stalls, sit at cafés, wait at *lavanderias*, park cars. Trees lean, a canopy of leaves covers the street, and through the leaves light carves waving white patterns into black asphalt. My phone buzzes in my hand, a text from Jay: *I think he's sorry. Or embarrassed.* A man sits near me on the bus; light flashes and he squints, closing his eyes against the glare. His gray hair goes silver then white then gray again as we pass below a sputtering sun. He's wearing a suit and I think to tap my opera ticket against my palm, alerting him to the possibility that we might be traveling to the same destination.

A bend in the street takes us into shade long enough for the man's eyes to reopen and there it is, that familiar hue, a dark and brilliant blue, and I'm struck with the urge to recite my accomplishments, assuring him I'd leave Rome in the morning having completed the Rome checklist: yes, I saw some steps and fountains and, yes, I ate the pasta and, yes, I was going to an Italian opera and, yes, of course I'd seen the Berninis, and wasn't this good? Hadn't I done well? At the next stop, the man stands. I listen to his breath as he lifts himself from the seat. His footsteps fade, he's gone, and me: I'd never stop wanting a father.

Off the bus, I follow a trickle of people through the streets until I see the signs guiding me toward *Nabucco*. I show a guard my ticket and am beckoned through a gate and into the gardens surrounding the ruins of the Baths of Caracalla. Above me, spotlights illuminate the underbelly of the parasol pines. The effect lines their branches in a silver glow, lightning stamped on the sky. Before me stands the skeleton of a grand imperial bath complex, one of Rome's seven wonders, the ancient *thermae*, where people once bathed together, talked, played, read. Vitruvius had his hand in this, too—his rulers and his calculations are everywhere, following me. He wrote a chapter in *De architectura* on proper bath design and Caracalla was built in accordance with it. The baths had lived many lives, first built in the two hundreds AD, then abandoned three hundred years later in the Gothic Wars. In disuse, it became a burial site, then a quarry, then a vineyard, and now concerts and operas were held here in the summer, and this is where I would see *Nabucco*, performed outside on a stage framed by the remains, by crumbling marble, brick, lime, and tuff.

My assigned seat is low enough that my feet can rest on the ground, a rare thing, and shallow enough that I can sit flush against the backrest, a rarer thing. I have an anxious feeling like I've forgotten something I'll need, and I stare out ahead of me into nothing until I realize the uneasiness is just surprise that I'm comfortable. Nothing about this assigned seat sends pain to any part of my body. Usually, chairs are too soft and pull my hips out of alignment, causing me pain. Or the seat is too hard and puts pressure on my spine, causing me pain. If I choose to sit back in a chair to support my spine, my dangling legs, too short to reach the ground, go numb within minutes, then throb and ache. For relief, I'll sit forward with my feet on the ground until my back is so sore that I must switch positions again. Surface, distance, absence, tugs uncomfortably on all bodies; my height, my curved spine, my mismatched hips only magnify the most commonplace pain. Discomfort skews my body language and people misread me. I fidget. I often cross my arms over my chest for added support for my frame, a gesture that can be mistaken for a signal of disapproval. Once, at a dinner party, a friend watched me shift and squirm in search of a reprieve. *If you're so unhappy here*, she'd said, *you can just leave.*

Jay and I had gone through a few rounds of negotiation as to where and how to sit when we'd met three months ago at the bar in Brooklyn. Anticipation scrambled my read of our email exchange and I'd accidentally arrived an hour early. I first stood at the bar, scrutinizing the menu, memorizing the menu; I stared at it until it knew my name. I would have liked Jay to have discovered me like this, my back turned to the door, gently tracing a line down the laminate surface with my finger, thinking. I imagined him then coming through the door, eyes scanning, finding me. The thought of him watching from the doorway made my chest constrict. I cared about this friendship, and it was passing through that early, precarious stage that ended (sometimes abruptly) with new friends choosing to become *real* friends or to politely stay acquaintances. I wanted the former.

I affected a casual lean. I could not pull this off. I ordered a cocktail I planned to nurse but saw my error as soon as the bartender walked it toward me. Its melting ice would mark for Jay precisely how long I'd been waiting. I drank it down.

Jay was a year ahead of me in our philosophy PhD program. We didn't know each other, really. We'd exchanged a few hellos in the hallways and polite nods while waiting our turn to fill up plastic cups with wine at colloquia receptions. At first, I'd barely noticed him, but his presence grew in my mind like shadows grow with changing light. Over time, the contrast between him and his surroundings was so sharp that my eye would find him first, even in a crowded room. Small things, an accumulation of gestures, brought forth this change. I liked the way he tilted his head suspiciously like a cat when listening to a lecture. I liked that he flicked his eyes down at his phone when someone made an inane comment in class so that they wouldn't see him grimace. Most special of all, perhaps, was the tone he used when asking questions. He spoke with a probing energy and genuine interest that—even when critical—had the effect of graciously opening a wider conversational space. Whomever he questioned was drawn into that space alongside Jay and, in this way, debate could be a collaboration. This was rare in our graduate department where most—students and professors alike—treated discussions as moves in the game of Proof I'm Smart. I liked Jay's laugh and his clothes and the way he spoke to women, which was the same way he spoke to men. From this collection of indications came a certainty: we would be friends. I knew he felt the same about me. Here's how I knew. One day, we were sitting on opposite sides of a large lecture hall. Two of our peers kept interrupting the professor to ruthlessly one-up each other and Jay had watched this skirmish with interest for a time and then grew bored—politely, then eventually less politely. He'd turned his back on the room and was staring out the window onto Fifth Avenue where a mime was humping a post office drop box. I knew before he did it that he'd turn, eyes skimming the crowd, to find me, and he knew before he turned around that I was already looking out the same window at the same thing he was looking at.

Jay was about to graduate and go out on the job market. It was too late for us to be real friends. I'd not made a proper effort and neither had he and that's the way of it sometimes. But then one evening, I passed Jay outside our school. He stopped to say hello. It had been raining and I dug my hands deep into the pockets of my raincoat.

"Oh," said Jay, "where did that come from?"

In my outstretched hand was a mandarin orange. I'd been surprised to find it.

"Ah, no thanks," said Jay.

I produced a second mandarin orange from my other coat pocket. Jay declined it.

"I heard you got a full-time job," he said.

"I did," I said.

"That was fast. I didn't think you were even on the market yet."

"I wasn't."

"Oh god, another one," said Jay, shrinking from my hand that held a third, small, fragrant mandarin orange. I opened my backpack. There were six more oranges inside.

"This is a strange magic trick," said Jay.

"Do you want one?" I asked. "I seem to have too many."

"Why all these oranges?"

"Can I offer you one?"

"It's a no-thank-you."

"Because you thought it was my only orange? I have more."

"Is this, like, a mom thing? To be loaded down with, uh, snacks?"

I felt a pang of separation at "mom thing." Was that all he saw?

I shrugged, he fled; he flung behind him a goodbye and was gone. I sank down into disappointment. I was out of rhythm with the whole philosophy department—its people, ideas, goals, strange language; their muscular analytics; their Grice, Quine, Frankfurt, Foot, and Fine. I'd wanted to study Plotinus, which was so pointless that I'm not sure I ever told anyone. I didn't have reasons for studying Plotinus, the Neoplatonist, nor an application for my knowledge of Plotinus. Worse, reading Plotinus sparked my most indulgent romantic ideals about philosophy—that it was good because it made you feel something about the big mysteries of human nature and the mind and the universe, and if you don't know how idiotic this is, just find an analytic philosopher and tell them you'd like philosophy to make you feel something about the mysteries of the universe. Correctly, no one took me seriously. People changed the subject when I walked up. I'd been told by a classmate that most of my cohort assumed I had been accepted to the program thanks to a diversity mandate. I'd never put anything in my application materials about my identity markers, but I didn't set straight the classmate who made this claim. I, myself, believed I'd jumped through some loophole.

But Jay—I don't think he'd ever thought of me this way. I think he respected me, even liked me. My entire career in the graduate program might have been different if I'd made friends with him from the start. But I'd missed that chance altogether and now this bizarre too-many-oranges encounter, which was likely to be our last and, well, now I'd messed up my chance to just fade neutrally from his memory. He'd graduate, get a job in Missouri, and that would be that. I imagined the future fully written. But then, an hour later, reality rewrote it. In my inbox, an email from Jay. It read, *Hey, Chloé! It was nice to see you earlier. We should get a drink or something in the near future if you're up for it.*

The sun sinks at Caracalla. A group of friends arrive together and take their place in the row in front of me. They settle in quickly, sprawling their bodies across their own chairs and those ahead of them. They wave and talk to people walking past. They are dressed in expensive flowing fabrics, both elegant and easy, as if this is a glamorous picnic, which, I suppose, it is. I'm overdressed, stuffy and formal.

This group of friends looks happy. Plastic cups and bottles of wine materialize among them. I watch one wrench his elbow, tongue out, opening the bottle of wine with a comic fiendishness. The man sees me watching and smiles. He says something to me in Italian. I smile apologetically. He moves the bottle from friend to friend, filling their plastic cups until the dark liquid trembles at the brim, making the person holding it laugh and scold him and swoop down quickly for a sip. I feel a wave of sadness watching these too-full cups, but I can't place why. I'm frowning now; the man in front of me notices and says something to me again and two others turn and look. The man fills a new cup with wine, this time not so full, and reaches it toward me. I can't understand what he's saying, but he keeps smiling and raising his eyebrows and tilting his eyes upward, I think toward the sky as if to tell me to pay attention to where we are, under the cover of perfect night, and so I take the cup and I breathe in deeply and I look up at the emerging stars. The smell of the wine is enough to loosen a knot in me. I take my first sip. It sends a shiver through, relief.

The sun pulls the temperature down with it and a cool breeze comes through, bringing the scent of the pines. There's no city noise here, only people's laughter and the creak of seats being claimed. The stage below is completely bare, just a black platform floating between the two crumbling pillars. I drink my wine and feel something heavy in me shift. For a moment I'm happy. The last of the sun glows through the arches of the ruins; rays illuminate the stage in layers, and I remember all the purposes these ruins have served. Time is not linear but stratified here.

I hear a sound and my body clenches. I know it immediately. Someone is laughing at me. I look back to the group of friends in front of me. They are all staring. The man with the wine bottle is standing in front of me with a strange look on his face.

He says, this time in English, "You like the wine?"

"Yes, thank you," I say. Had I not thanked him before? My face flushes. I begin to sweat. My cup is empty now and he moves the bottle toward me.

"More for you?" he says. I smile and hold my cup toward him. His friends erupt. They turn their faces into each other's shoulders.

"Maybe," the man says, "first you do what I asked before?"

It's then that I feel eyes on me from above and below. The man's friends are all looking at me, but it's not just them. I turn and behind me is another couple. They are watching me closely, watching the man and me.

"Oh," I say; a picture is forming, my mistake comes into view. "I'm sorry. I'm so sorry."

"It's OK," says the man, smiling. "But this time you'll pass the cup back to my friends instead of keeping it for yourself?"

I sat in a booth in the bar in Brooklyn waiting for Jay. The booth was too low to the ground, which made the table seem disproportionately tall. The seat was too soft, and I sank down into it, my chin just visible above the table. I was grimacing by the time he arrived, straining to prop myself up by my elbows in a failing attempt to look like a casual person in the act of sitting.

Jay arrived at the bar at our agreed-upon time to the minute. I watched him for a moment before he saw me. He ran his hand through his curly hair and smoothed his jacket. The precision of his punctuality opened the possibility that he'd also been nervous and had arrived early like me, maybe pacing inside the bodega on the corner until it was time to meet me. He squinted at me in the booth and pointed to a low table with hard wooden stools. This table was pushed into a crowded corner. There was more space at the bar, but Jay shook his head at that possibility and pointed again to the table.

"Right?" he said. "That one is better?" He looked at me a bit unsure. "For you?"

We moved. This new table was the inverse of the booth, too low and the stools too high; they pinned my legs beneath the tabletop. It was painful and I reflexively pulled my shirt above my mouth, a conspicuous but ingrained habit, to hide my grimace.

That night, my body was a heavy thing between us, and I saw it reflected back to me in the tender, anxious way Jay processed the problem of where to sit. He was trying to keep control of his eyes, trying to tame the way they flitted over me with concern. He looked across the room, to the tables, and back to me. I knew his worry came from a stab at kindness, a well-intentioned attempt I was supposed to accept with a smile. I put one on, a muzzle. I just wanted the spotlight on my body to lower. I wanted to be a normal person having a normal drink with a new friend.

Jay took a prescription vial from his pocket, shook out a pill, and swallowed it with a mouthful of beer. Behind him, the bar's windows framed waning light. The room was dim, but I could see Jay's face clearly by the glow of the small candle lit between us. Jay, tall and broad, hunched over the tiny table. His movements were slow, deliberate; his laugh, a foghorn. The authority of his presence distracted me from his piercing eye. He had a bead on everyone in the bar, especially me; he saw into my muted frustration, my embarrassment over the problem of where to sit, and I understood that the pill bottle he'd brought out was an attempt to connect with me.

"Bomb in the brain," Jay said, shaking his pill bottle.

"Bomb?" I said.

"Yeah, that's right," he said. "A tumor. If it grows, I die, if it stays the same size, I'm fine. But these pills have kept it stable for a long time."

"But if it gets bigger, you'll die?"

"Yeah."

"You're very nonchalant about it."

"What are you going to do, worry all the time about death?"

"That's my primary mode."

"Are you in imminent risk of dying?"

"No more than most."

"So why?"

"When I was younger, there was some fear that as I grew, my body wouldn't be able to support my organs."

"Are you technically a dwarf?"

"I guess so," I said.

"You are, by definition."

"I'm short, it's true."

"Let's look it up," and Jay looked up the definition of *dwarf* on his phone.

It was then I noticed a new body watching us. It stood hesitantly in a doorframe.

Jay waved the figure over to our table and said to me, "I hope you don't mind."

I didn't recognize the man at first. He introduced himself as Colin and told me he was also in our philosophy program, a year ahead of me. Colin bought a round and settled into our tiny table. He'd just finished teaching an evening ethics course at a CUNY school nearby. He began to complain about a student. His voice rose and Jay reminded him he ought to keep a low profile.

"I know," Colin said and he took a long gulp of his beer. He stared down into his drink, then said mournfully, "Why did we have to come to this bar?"

"Is there a problem?" I asked.

Yes, there was a problem. This bar had been the site of *an incident*. Jay began to explain.

"First, you have to understand," said Colin, interrupting Jay. "I'm angry all the time."

Jay continued. One night, months ago, Colin had come to this bar, opened a tab, bought a few drinks. Jay bought him a few more, and then Colin was drunk. He went to the bar and, forgetting his first tab, started a second. He realized his error at the end of the night and tried to get both tabs closed and both credit cards returned, but it was a dark bar and a loud bar and, well, this bartender didn't fully understand what Colin was asking for because Colin was very drunk and wasn't speaking clearly but he kept trying to make himself understood and the bartender kept mishearing and this was when Colin became very angry and he'd yelled, *You are a fucking bitch.*

"Wait," I said, feeling I've somehow missed relevant parts of the story. "Why did you—"

"The Asian one," Colin said. "Very young. She's not here tonight." He tipped his glass and the last of his first beer disappeared down his throat.

I looked at Colin and then at Jay. This story was not an ideal introduction to Colin.

"She didn't ask you which bartender," said Jay. "She's trying to understand what happened."

"Oh, I don't know," Colin said. "I felt terrible about it later." He'd yelled at this very young, maybe Asian, bartender, *You are a fucking bitch*, over and over until eventually a bouncer came for him.

"Colin had just started a new medication," said Jay. "The new meds made him angry and paranoid."

"And remember I told you," said Colin, "I'm already angry all the time."

No one said anything for a while, then Colin shouted, "So what? I'm depressed! So is Jay. So is everyone."

Jay shrugged and nodded, and they discussed Colin's previous depression medications and agreed his current ones were much better. Jay had tried many of the same meds and therapies. The two men went back and forth, tracing their shared histories with pills. They got more drinks. They rattled off names of drugs I recognized only from television commercials.

The more Colin and Jay talked, the more evident it became that they knew a lot about each other. This surprised me. They did not strike me as people who could be close friends. They were opposites in tone and temperament. Colin stumbled into modes of being. He drank like he was being timed. He was trapped in his moods, tossed about by the torrent of words flooding from his open mouth. When he was complaining about a student, he was red with rage. When he was talking about his sick dog, he was in despair. There was a logical side of him that was constantly doing battle with the part of him that spoke. He would say something and then throw his hands up as if to say, *Oooh, this guy is the worst*, referring to himself.

Jay was more calculating, self-possessed. He was aware of his effect on a room. He and Colin both dressed like shlubby grad students, but with slight variations. Colin dressed to appear as the "shlubby-grad-student type" whereas Jay dressed to be a particular instance of shlubby grad student. The difference was Colin wore black Converse and Jay's were an aggressive chartreuse. I liked Jay, but I knew he was less likely to tell me the truth if the truth didn't reflect well on him; I didn't like Colin but saw that his lack of guile meant he was, in his way, more honest.

Whatever their differences, they were united in their understanding of each other's depression, which, they told me, required thoughtful, constant vigilance.

Colin said to me, "Imagine a cup that is filled as far as it will go."

Jay nodded to indicate they'd used this analogy to explain themselves to others before.

"Imagine water at the top of the cup," Colin continued. "Imagine a meniscus. There's a delicate surface tension required to keep all the water in the glass. Sometimes I feel like that. Completely filled with sadness or anger, filled dangerously to the brim."

"Some days," Jay said, taking over, "I wake up and the cup is full and if I get out of bed or go to school or talk to anyone, that cup has to come with me. I feel it always. I feel it between us now."

"Any drop added will break the surface tension and the water will spill," Colin said. "Does this make sense? That's what happened with the bartender. It isn't all the time. Just sometimes. Do you understand? Can you understand?"

I did understand and I didn't. We were not the same, we did not have the same experience, but I, too, lived in constant negotiation with a different too-full cup.

Sometimes I wake so full of pain that I wonder if I can get out of bed. Every move, every weight, every step or stair, every block I walk adds up, ever approaching a limit. I'm constantly trying to consolidate the tasks that will require my body to be upright and in motion. A part of my mind is permanently lost, swallowed up by endless pain calculations.

As Colin and Jay talked, I felt a thrill; there was a chance at being understood and I saw the possibility of a new kind of closeness.

"I'm angry," said Colin. "And the wrong medication makes me angrier."

Jay said, "But no medication at all—"

"We're dead," said one man as the other said, "Oh, we're dead."

They explained themselves with the frankness of shared knowledge. Their lives overlapped enough to confirm the other's reality. In each other they found recognition and from there they'd formed a bond.

Plotinus wrote our soul has a special faculty for the recognition of beauty, that the tingling feeling of excitement in the presence of beauty is our soul recognizing itself in an object. This is kinship. *Any trace of that kinship, thrills with an immediate delight, takes its own to itself, and thus stirs anew to the sense of its nature and of all its affinity.* I see the beauty of kinship crackle between the two men as they speak. I'm outside of it but maybe only a little. I never talked about my disability with others because it immediately put us on unequal ground. My silence kept me separate. I shared myself with no one. I waited for the new people I met to unsee my body, to forget to stare, which happened over time. Enough exposure to me dulled my effect. Each person I met took their own amount of time to forget. And I could be patient.

As I listened to these men discuss their depression I began to wonder if I could feel with them this kinship borne of recognition. They said things I understood. They, like me, stayed vigilant; their pain was easily transferred to others—to strangers, bartenders, students, or, most easily, to the people they loved. I imagined the possibility that these men could both be a new kind of friend to me, that we could be, maybe, more empathetic to each other than not. Desperation for this alliance helped me unhear some of Colin's uglier utterances. I felt their circle widening, approaching me, and I saw my place within it, and I was glad. But then Jay and Colin kept talking.

Plotinus, who lived in the third century CE, did not want his body. Even his closest friends and students claimed to know nothing about his family, age, or where he was from. Those facts spoke to corporeal realities, which Plotinus dismissed as damaging distractions.

Much of Plotinus extends from Plato's theory of forms and his realm of perfect ideals. But Plotinus believed the realm of ideals held not just the archetypal form of love, beauty, justice, bed, chair, or couch, but it held the archetypal version of you, the individual person. Ascending through the mind to the realm of ideals allowed you a glimpse of your perfect self. That perfect self was only a mind, separate from a body.

Plotinus, like the philosophers who follow from his tradition, lionizes the endless and impossible quest. The quest for the Good, the True, the Beautiful, can only be undertaken if first we divest ourselves of the filth we've acquired during our time in our bodies. A Supreme and Eternal Beauty awaits those willing to cleave mind from body.

Plotinus: *Your body is the problem, get rid of it.*

Detach from it.

Be a mind, alone.

Purify and put aside your garments and enter, naked, into the Holy Celebrations of the Mysteries. Remain in solitude, in the Solitary Existence, the Apart, the Unmingled, the Pure.

The boys got more beers. The conversation returned to Colin's annoyance with his students, which was really an annoyance with adjunct labor and the thin academic job market. In what I intended as an act of commiseration, I recounted the story of Steven.

I'd been teaching biomedical ethics, a field that applied ethical principles to dilemmas pertaining to the body. I often used real case studies to illustrate certain ethical principles. That day I'd taught a case I hated teaching, but never skipped.

The scenario goes like this: A married couple, who cannot conceive on their own, uses IVF to become pregnant. They successfully fertilize four embryos. The couple is deaf, and two of the four embryos are also deaf, while two are hearing. They want to share their culture, their language, their lives with a deaf child. Are there any ethical principles that would prevent them from selecting only the deaf embryos to insert in the uterus?

I taught this case every semester as a form of punishment for myself with little consideration as to why I felt I deserved to be punished. Reactions ranging from discomfort to disgust rippled across the faces of some of my students as I read this case aloud. I dreaded the arrival of this moment in which I'd be standing alone in front of a room, reading reactions, seeing which prejudices might be betrayed. But I also longed for this moment, looked forward to it and cherished it. It was like a haze cleared. I liked seeing their masks of goodwill dissolve. Too often, my students treated me as if I was theirs to care for. "Take your time," one student was fond of cooing to me as I limped my way to the front of the room each morning. "Be careful," she'd say, "no need to rush."

I asked for my students' assessment of the case, but no one wanted to start. They did not like my watchful eye. They were comfortable applying their gaze to me, less so in the reversed sightline. I absorbed their discomfort. Their resistance exhilarated me. That the case was about deafness and that I wasn't deaf did not matter to them—for most of my students, the deaf and I were inscribed within an imaginary circle of damaged people. I read this case aloud, I kept my voice clear and steady, I waited, I watched; it was a dare, a chance for people to reveal themselves.

The field of bioethics is grounded in four principles: respect for an individual's autonomy; a fair and just distribution of resources; beneficence, which is the commitment to assuaging suffering; and nonmaleficence, which is the promise to do no harm. On a good day, a medical professional could honor all four of those principles in every choice they made. But often two or more principles came into conflict with each other and that's where the most interesting, complex biomedical ethical cases arose.

Some students understood immediately that the deafness case was about autonomy. Those students easily rattled off consistent, principled arguments as to whether or not any parents should be allowed the autonomy to test and select embryos based on any genetic factors. They saw no principled difference between testing for deafness and testing for any other genetic trait. But most others in the class allowed their arguments to be led by their visceral, gut-level, negative reaction to the choice to bring a deaf child into the world. A student said autonomy is good, but this is too far.

"Too far from what?" I'd asked.

"From what's normal," the student said.

Another student raised her hand and said, "It's cruel for the parents to force a child to be deaf. Doesn't cruelty to a child violate the principle of non-maleficence?"

"So, I waited," I said to Colin and Jay. "I let that comment hang in the air until finally another student struck the argument down."

"Struck it down how?" said Colin. His tone should have been my first clue, but I missed it.

"By reminding the class that pre-implantation genetic tests revealed existing realities of the embryo. Selection doesn't change the embryo or 'force it to be deaf' any more than one could force an XX chromosome to transform to an XY."

"Did this convince the first student?" Jay asked.

"No, she was set in her belief that deafness was a cruelty." I rolled my eyes. I continued. "She responded by saying something like—if you can believe this—that it was just unethical to knowingly have a deaf child. She was like, *Their lives will be worse than the normal child!* Normal. *Normal.*"

"Ah," said Jay. Colin said nothing.

In class, the argument between the two students had continued:

One said, "Could you say to the parents: OK, you have to choose the male-sexed embryo and it is unethical to choose a female-sexed embryo because she may have a more difficult life in a patriarchal society?"

"No, because that's not necessarily true," said the other.

The first said, "But it is necessarily true that a deaf person will always have a worse life than a hearing person?"

"Well, I wouldn't want to be deaf," said the other, meaning to end the argument.

"That's fine," her opponent pushed on, "but if you allow the parents the autonomy to take the tests, then force them to choose a hearing embryo and discard the deaf ones, you are saying that hearing people are inherently more deserving of life than people who can't hear. Do you see a problem with that?"

At this point another student raised his hand. Steven. He was my favorite, funny and thoughtful in his writing, quiet in class. He drew in a sketchbook throughout the hour. I called on him.

"It's unethical," Steven said, "because it's dangerous to be deaf or disabled. So, the parents are willingly putting their child in danger."

"What do you think is dangerous about being deaf?" I asked.

"Think about it," he said. "Deaf people can't cross the street by themselves safely if they can't hear the traffic."

The room was silent. The two students who had previously been arguing seemed glad to have the attention off them for a moment. The class waited for me to speak. I paused here in my story and looked at Jay and Colin to try to read their reactions. They listened to me, faces blank.

"How do you get to school?" I'd asked Steven.

"I take the train," he said.

"And then do you walk the rest of the way?"

"Yeah."

"So, you cross a lot of streets, some which are very busy, crowded streets."

"Yeah."

"Like Flatbush."

"Yeah."

"And Bedford."

"Yeah."

"Busy streets, lots of people, cars, buses."

"Uh-huh."

"And do you ever wear headphones?"

Steven looked at me for a minute, then he broke into a big smile.

"Every day. Every single day," he said, and he stood up, walked to the front of the room, and offered me a high five, which I accepted, cringing.

"You are right. I can't hear shit when I have my headphones on and it's totally fine," he said, laughing at himself. The class laughed, too. "I don't think a thing of it."

"Yeah," I said.

"Wow, OK, I was wrong. Why did I think that?"

"Because you weren't thinking," I'd said and left it at that, but what I thought was, *Because you weren't thinking of a deaf person as a whole person.*

With my story over, I watched Colin and Jay. I smiled. I still believed I was among friends.

"Right," Colin finally said. "Well, that's bullshit."

Jay looked up from his beer.

"They should be locked up, the parents, if they do allow a deaf kid to be born on purpose," said Colin. "All pregnant women should be legally obligated to undergo testing for any kind of disability and if they find anything, they should be forced to abort. And if they don't, they should go to prison or have to pay a fine or something."

"I don't think you mean that," said Jay. He proceeded to calmly argue the logistics of Colin's proposal, pushing Colin to lay out his necessary and sufficient conditions for "disabled" in this context.

"What you are arguing for," said Jay, "is a return to eugenics. That's what you want?"

"Yes!" Colin clapped his hands. "It was a great idea, one with true ethical merit, it's just unpopular to admit it."

Their debate went on without me. I tunneled within myself until I arrived at my familiar place, the neutral room within my mind.

Dr. Asher, my orthopedic surgeon, had been the person to teach me about the neutral room. He monitored my spinal abnormalities for many years, so we saw him often and I knew him well. I was young when we had our first conversation about it. I remember him asking me questions about the level of daily discomfort I experienced. I was using crutches then to walk and I had braces on my legs. My mother had left the exam room and was waiting for me outside and so, because we were alone, I could be honest with Dr. Asher. The discomfort I felt was more than physical pain. I worried I was slowing my mother down, making our everyday tasks more tedious. I dreaded doing the shopping with my mother because it involved so much walking. I panicked when we couldn't find close parking and, while my mother looked for a spot, I'd start to spiral inside, knowing I'd soon have to cross a long lot, then walk the long aisles of a freezing grocery store, then wait in long lines. My mother never complained. I think she was always happy to have me with her. But I saw how hard she worked, how little help she got, and I loved her so much. I did not want to be her burden. This is the most visceral memory of my childhood, this anxiety I felt, being in the car with my mother as we circled these huge Midwestern parking lots.

Dr. Asher listened to me. He told me that imagining the pain of future events could trigger pain in the body in the present. My mind and memory were powerful, he taught me, and could be used to quiet pain rather than amplify it. He said, *Stop thinking of the grocery aisles and lines. Instead, focus on a car a few yards ahead in the parking lot. You are only walking to that car. Take a deep breath, go to a controlled place in your mind. There is no parking lot, only that car, about eight steps away. Once you pass that car, look for the next landmark, eight steps away. You are only walking for eight steps. Count the steps.*

The neutral room made pain easier to manage. It lasted only for eight seconds at a time. Other things started to become easier, too. All difficulties could be broken down into discrete, achievable tasks. School stopped feeling overwhelming and I excelled. I took a deep breath, went to the calm, enclosed room in my mind, and focused only on the next paragraph to read, the next question on a test. There were comforting certainties in the neutral room: I could hide there, numbed and patient. *Where do you go to find escape from a reality that is oppressive to the soul?* I went to my neutral room where time passed in counts of eight, and the distance between present pain and future relief was objective and measurable.

But sometimes, when the pain I was experiencing was especially intense, I'd retreat so far into the neutral room that I dissociated a bit from reality. I'd be looking out my window and then suddenly it would be dark and I'd realize I had no sense of the hours as they passed. I could sit in lectures or listen to people talk at parties and not retain a single word. I lost objects constantly. Andrew had once called me "the Michael Jordan of losing an iPhone." I never knew where anything was. When we started dating, Andrew had said, *The tactile world is less real to you than it is to other people.* The neutral room had contributed to our earliest fights, before he really knew me, because there were times when he would be standing in front of me, talking to me, asking me questions, and I would not see that he was there at all.

Colin and Jay talked on and on and, at first, I chose not to hear them. But slowly I began to feel something new: an infection, a heat. It was anger, and it forced its way into my closed room, distorted and disturbed it.

My student, Steven, had laughed and said, in giddy self-disbelief, *Why did I think that?* I'd laughed, too. I'd let him off the hook. How many people felt my life was inherently worth less than theirs? How many people did I meet who felt the same way Colin felt about me but were just too sober to say so?

How much of that was I responsible for?

I heard Colin's voice and I sensed myself being pulled from the safety of my retracted mental space. His words landed on me, I started listening. My sensations doubled, I was awake, I pushed up my sleeve, exposed my skin, traced my flesh with the edge of my nail, smiled.

"You know that I'm disabled?"

"Yes," said Colin.

"Do you think I shouldn't have been born?" I asked.

"You're born," Colin said.

"But in an ideal world, I would have been discovered and aborted?"

"Yes," he said. "Your body makes your life worse, harder. This is just an obvious truth."

"You think my whole life, every aspect of it, is made worse?"

"Do you deny that? I don't think *that's* the controversial statement here."

"I think it might be one of them, yeah," said Jay.

"Does it surprise you if I say that actually I think my differences have brought many positive things into my life?" I asked.

"It doesn't surprise me that you'd say that," said Colin. "It's the same thing deaf people mean when they talk about deaf culture. But it isn't a culture, it's coping."

"That's not a supportable position," Jay said.

"Look," Colin said to me, wearing a wounded look, "I'm sorry if I've offended you."

"She's not telling you you've offended her," said Jay. "She's telling you you're wrong."

"I don't mean to hurt your feelings," said Colin.

"She's not saying you've hurt her feelings!" Jay shouted. "She's saying your claims are not factually correct. Do you see the difference, Colin? The difference is important."

"But if you're honest," Colin continued, "when you say disability could have a positive aspect to it, you're just—and I understand why you'd have to do this—you're just inventing a narrative in which your life isn't worse than other people's. And listen, we all must tell ourselves a story in which we're, like, wow, so grateful for struggle and suffering. It is nice to think resistance is strength or whatever, but that's all made up, it's all rationalizing. You're rationalizing the shitty hand you've been dealt."

He turned to Jay and asked him, "If you could just snap your fingers and never feel depression ever again, wouldn't you do it?" And Jay agreed he would. "You see," Colin said to me, "depression has made my whole life worse, every part of it. And that's the truth. There's no spin. I want to die." Colin pointed at Jay. "He wants to die sometimes, too. We're on the job market now and what if we can't get tenure-track jobs? After all these years working toward the PhD? If it all amounted to nothing? I'd probably just have to kill myself."

I looked at Jay, who was nodding.

"But," said Colin, "at least I'm not disabled."

So, there it was. He saw a body like mine and thought: *Things could be worse.*

Later, I stood on the street with Jay. I'd hear in the following week that Colin would recount the night to other people in the philosophy department, apologizing to them as if they were my proxies. A few would report back to me that Colin felt embarrassed about what he'd said, that he'd been swept up in a drunken stubbornness that masked whatever it was that he'd really meant to say to me.

"Can I walk you home?" Jay asked.

Franklin Avenue buzzed with bodies as the bars expelled their patrons. I felt a drop of rain and looked up, blinked, and saw it was not rain but an air-conditioning unit dripping on me. I heard a metallic rustling that first sounded like the wind chime hanging on a post on my mother's porch. The sound I heard was coming from an approaching man whose outstretched arm slithered through the air, tracing something I couldn't discern until he walked under the streetlight. Silver sparked around his lifted hand, he was shaking a tambourine. He walked toward us, singing. On his shirt in big block letters: I NEVER FINISH ANYTHI

"How do you feel?" Jay asked.

"Complicit."

"Why?"

I had no language in this moment and I didn't know if it was because I was exhausted by the work of explaining myself or if I couldn't explain myself.

We reached my apartment and Jay stood in the doorway. "You know what I kept thinking throughout that whole conversation?" he said. "I kept thinking that here I am, depressed, anxious, tumor in my head, other shit, but no one looks at me as disabled. But you . . . you can never not be seen as . . . What separates us, really, when you think about it?"

Jay had looked uncomfortable all night sitting at the short table. He hadn't even attempted to fold his long legs below the tabletop, but instead had slid them around the edges in an awkward straddle. I, two feet shorter, had also been uncomfortable at that table. All night, we'd done our best to adjust; me in my ill-fitting body and he in his ill-fitting body. Tables and chairs were not made for either of us and we knew this and did the best we could, choosing to sit together anyway, to drink, to talk, to lean awkwardly, painfully, toward each other.

"Are you going to write about disability?" Jay asked.

"No," I said. I turned to my apartment door. I wanted the night over with, not because I wanted to get away from Jay, but because I didn't want to hear the rest of this conversation, I didn't want to hear what I would say.

"Did you ever write about disability?"

"No," I said.

"Why not?"

"I'm tired."

"Did you ever want to write about—"

"No," I said.

"Because?"

"I just . . ." I didn't know how to finish that sentence.

The first time I'd ever seen Jay had been at a welcome party for my cohort in our first week of our PhD program. I didn't know anyone. I was nervous and had a stronger-than-usual urge to hide my body. I'd given birth to Wolfgang six months prior. My breasts were heavy with milk and I had wads of toilet paper stuck in my bra to soak up any leaking. I needed to go home, but I also needed to be in this room with these other philosophers, one of whom would give me a sign that I'd made the right choice, that there was something here for me, in this new department, in this new life, in this huge city we'd left Kansas to live in. A group of students stood near me. Some in the group were new like me and some were not. They were gauging interest among each other in starting a philosophy of perception reading group. They talked about what to read and when and how often to meet. I gathered my courage. I took a step toward this group, opened my mouth, but then I took a step back. I was alternately afraid and then furious with myself for being afraid. I thought of Wolfgang, of how holding him felt like hugging a sticky bag of flour. My milk surged. It would show through my shirt soon. I needed to go home, back to our apartment, but I couldn't because I'd moved across the country to join philosophy of perception reading groups, to meet at bars and talk to philosophers; this was, idiotically, the thing I'd sacrificed so much to do, but was also the thing I was most terrified of doing.

Finally, I said to the girl closest to me, "Are you forming a reading group?"

"Yeah," she said. "Are you interested?"

"Yes," I said, blushing with relief.

"It'll be pretty intense, like, for people who are looking to write their dissertations in the field. It won't be social identity stuff."

"What do you mean?" I said.

"Do you focus on, like, body and identity theory?"

"No," I said.

"Oh," she said quickly. "I really thought that's what you told me you were studying."

I'd never talked to her before.

"No," I said.

"No, I swear that's what you said."

Jay had been standing in that group, but he hadn't noticed me, he'd looked right through me.

I watched Jay outside my bedroom window. He lit a cigarette. He stood there smoking, staring out into the street, then got into a cab and was gone.

Andrew was at the very edge of our bed, his back hunched in sleep. Wolfgang was sprawled horizontally across the rest of the bed, his covers kicked to the floor. We could never keep the blankets on him. Even on the coldest nights he flung them from his body. I kissed his wet and sweating face, listened to the rhythm of his breath, and replaced his blankets, tucking them under him. He stirred, kicked his legs, and freed himself.

The oranges I'd offered to Jay when I'd seen him at school, the snacks, had not been a "mom thing," or it had been, but not in the way he'd implied. It was Wolfgang who'd filled my pockets and backpack. He'd heard my stomach growl while he'd clung to me in a long hug and he'd asked me what was wrong and I'd said that I'd forgotten to eat lunch. He'd snuck into the fridge and packed me the best lunch he could manage.

Wolfgang was always watching me, always worried. He was, it seemed to me, uncommonly sensitive. He'd sit in the kitchen while I cooked, but cried when I smashed garlic or sliced into the flesh of an apple, seeing these actions as violent. He believed walking on grass hurt Grass's feelings. Once my mother had drawn a face on a balloon and he'd carried that balloon with him everywhere he went, named it Potato, and wept when it popped. He had long, intimate conversations with his stuffed animals. He was constantly anxious about causing harm or harm coming to others. He wanted to be near me always. He searched my face for evidence of my discomfort and when he found it—a cringe or grimace—he cried. Even a sharp intake of breath could send him spiraling, needing to know what was wrong. There was no reassuring him. He didn't listen to what I said, only absorbed what I did. There was nothing I could hide from him. He sensed my moods and my rumbling thoughts, my empty belly, my every distraction, frustration, fear. He felt them all and carried them with him.

Earlier that night, before I'd gone to meet Jay and Colin at the bar, I'd been standing in the doorway of Wolfgang's room. He and Andrew had been playing on the floor, building something with blocks. Wolfgang waved one in my direction, an invitation to crouch down on the ground with them, but I couldn't move. I was stuck in the doorway, watching them as if we were separated by a glass wall. I couldn't join in. I couldn't cross the threshold of Wolfgang's room. I didn't belong there. I belonged elsewhere, apart, pinned at a remove. The streetlamp outside imposed a yellow hue onto Wolfgang's face. If I entered their scene, I would alter it, make it something else. I wasn't sure if it was me or the scene that should change.

Here I was, after the bar, in their doorway again, watching them sleeping. I counted their breaths, counted to eight; I needed to see eight deep breaths to believe they were real, that they were alive and OK and would be OK without me. I walked to the living room and lay down on the couch. I stared into the cracks on the ceiling, little spines spreading across an expanse of dull white.

I'd believed completely that it was my nature to exist at a distance, to be essentially, at my core, alone. Plotinus and me: Solitary in Existence, one of the Unmingled, the Apart. A husk had grown up around me bit by bit, building a boundary between myself and all else. It grew up from all the people staring, always staring, from all the times I moved through public spaces, greeted by whispers. *What's wrong with her? What's wrong with you?* It grew when I was kept separate from other kids at gym class or recess; it grew when I was excused from school field days; it grew when I attended anyway, not wanting to miss out, and was trailed by the eyes of my teachers, brimming with concern. Concern could turn to anger when people felt my presence was a reminder of their insensitivities, their unintentional exclusions, their failures to plan a curriculum that included me. The message was clear and it came from everywhere: I was not to participate. I was not a participant.

As I got older, I learned there were abstract spaces where I didn't belong either. Sexuality, romantic love, partnership—these were all part of a province I, as a disabled woman, was not meant to enter. I saw people cringe when I mentioned a crush or joined my girlfriends in lusting over a celebrity. I was neither expected to date nor marry. It was presented to me as a certainty that motherhood was another space I was excluded from. I accepted what I thought was a fact. And as a result, a part of my imagination never developed, the part that wonders what it might be like to love your own child.

I made people uncomfortable and sometimes they were cruel, but the much more common experience was that people simply felt it was hard to include me and easier to leave me on the margins. My body was constantly seen, but this thing I called my "self" was invisible. I learned to preempt the inevitable and excluded myself. I retreated to my solitary spaces before I could be pushed out of the flow of this realer life, this life that glistened all around me, bright and full and inaccessible.

My father had used travel, adventure, theory, philosophy, art as materials to build a great scaffolding around his theory of self—protecting, separating, and elevating it. I, too, did this, and told myself that I was made better by being separated from people. Some part of this was true, but I couldn't tell it from the less-true parts. I fell in love with philosophers who showed me how to erect grand theories of divine loneliness. I was Plotinus's disciple, determined to find my perfect self in the Solitary Existence, the Apart, the Unmingled. Beauty helped lift me out of the filth, beauty broke me free of reality and helped me be alone. The ancient Greeks tell you to yearn for *sophrosyne*—a moral sanity, a state of balance and harmony with reason and wisdom. For Plotinus, *sophrosyne* was achieved by shutting out the sensations of the body, both its pains and pleasures. The body is unclean, unworthy, and beauty is found in our divestment from it. *The Soul thus cleansed is all Idea and Reason, wholly free of body, intellective, entirely of that divine order from which the wellspring of Beauty rises and all the race of Beauty.*

I heard a cry from the bedroom. I ran. Wolfgang was sitting upright, lost in a bad dream. He was drenched in sweat, eyes open but not seeing, his head whipping from side to side. He mumbled a strange incantation. I took him gently in my arms. He closed his eyes and slept.

The lights go down at Caracalla. I can't make out my hand waving in front of my face. I can't see another person. The dark collects and settles us. The crowd clears its throat one last time, expelling a jangled tune, a song of scrapes, squeaking seats, murmurs, whispers. The hum and drone of these final noises rise and evaporate above. Then a hush. We wait for the next moment to arrive, but it hangs suspended and time expands instead of passes and I'm tense in attention at first, but then I, too, expand. My mind slides, my body unclenches, the cool night air and my little cup of stolen wine take full effect and I come loose. In the darkness, the clamorous world winnows away behind me. The only visible light is what comes streaming through holes punched by stars in the sky.

Softly, the stage lights rise. Performers appear, standing in slate-gray costumes. Their bodies form a wall. Dim light tricks the eye into thinking that there is no end to their line of bodies. A green light edges upward, revealing the looming ruins of Caracalla, and although I'd been staring at them for the past hour, they are renewed in this light, this new context, and the pillars now lend the authority of time—its ancient nature and this moment existing at once, collapsing and rising on the stage before me. No one moves.

And then!

The hundreds of bodies convulse and they sing out together and everything around me is transformed by the sound of them. They sing the same note, they sing the same phrase, they are—so the story of *Nabucco* goes—the Israelites praying, sending up a lament for their fate to an angry God. The Babylonian army advances on Jerusalem. The voices call out across the open space. An actor, just one actor in the mass, tips his head up to the sky as he sings; he is asking for God to intervene.

These unified voices throw a lance across the stage and it pierces me. The sound they make cleaves from me the dullness of the day. One sound, one note; pure, apart. This clean sound, these voices all singing together. It lifts me above the muck of the ordinary and holds me above my materiality, above my muddled perceptions, my gnawing need. It reminds me of the incredible pleasure of being separate and not among. For a moment, I become the only living thing in the world—the performers disappear, the crowd disappears, Rome disappears; cars and cafés and *lavanderias*, gone; New York City, gone; my new job is gone and the people who wait for me at home have vanished; my fears are gone. There had been a weight on me, a certainty that I'd been whittled away to nothing—by motherhood, by marriage, by work, by body, by pain—but I can't feel that weight anymore. There is no place for me with people. Art lifts me from the mud but *people* could not do that for each other. Every time I tried to join them, I was humiliated. They bring with them too many eyes, too many ideas.

I think then of Plotinus, but I also think of a favorite essay by the philosopher Iris Murdoch in which she'd written about beauty's transformative powers, arguing that an experience of beauty could change one's "arid consciousness" and could lift a person out of themselves, out of the world entirely. She says she sees a bird, a hovering kestrel, and "In a moment everything is altered" and she says something more but, because I am, in that moment in Rome, a hurt and brooding and lonely self, I do not correctly remember the point of Murdoch's story of seeing the kestrel. In a year, I'll read her again and I will remember this moment at the opera and my error in understanding her. But for now, her point is mine to badly interpret and I use it as evidence for what I want to believe, which is that I am better off alone, beholden to no one and nothing, free from dreadful normalcy, and I sit down into this feeling of divine loneliness. And I feel relief. A solitary future writes itself.

But, in an hour, it is rewritten.

The opera goes on, beyond that single note. Its story unfolds. The stage is stripped. There's nothing to dazzle the eye here, nearly no color, no bright lights. Just voices, music, the looming pillars, the surrounding line of stone pines. The grip of the opening note loosens, and I hear the breath of the people around me; I hear them begin to enjoy themselves, they pass between them cups and bottles of wine. It's a picnic again, a summer scene.

In the third act of *Nabucco*, the Italians in the audience all stand up. They know something I don't. In front of us, all the actors line the stage again, just as they had at the start of the opera. But now they are kept from us by a metal gate. They sing from behind this gate. Their arms stretch through it, toward us. We see them in fragments, a square of skin, a patch of fabric. They reach for us, the audience, and the audience reaches back. Everyone around me begins to sing together: *Va, pensiero, sull'ali dorate*. I feel arms lift me from my seat. It's the man whose cup of wine I mistook for my own. He's turned around to face me. He throws an arm around me and sings. Behind me, beside me, more hands reach out and grasp me. Waves of sound bounce back and forth off the walls of the ancient ruins. It's all, for a moment, colliding here, in compressed time. There is history braided into this moment. Verdi had thrown a libretto to the floor and there it opened to one line: *Va, pensiero, sull'ali dorate*. "Go, thoughts, on golden wings."

At the premiere of *Nabucco* in 1842, the audience demanded the performers sing "*Va, pensiero*" again, using it as a means of protest against the Austrian Empire and its prohibition on encores. This origin story is suspected to be mainly myth, but it doesn't matter what's true, it only matters what story gets written. "*Va, pensiero*" became a patriotic song, an anthem of the Risorgimento. In 2011, a conductor interrupted his own performance of *Nabucco* to give an impassioned speech about the value of art and then made everyone stand and sing "*Va, pensiero*" together and they did, throwing their playbills in the air and then letting them float in a soft swirl from the rafters.

The Italians around me are all standing and singing it now and I sing it with them even though I don't know the words. I stumble through, strange arms around me, strange hands gripping me, all of us in the middle of a sentimental and imperfect history, joining voices, singing together, and I am on the edge of a feeling, and my skin turns to gooseflesh, and I yield to the sensation of being touched, held, a part, among. In this collision of echoes, collapsing shadows, merged voices, human skin and sweat, there is a new beauty. The purity of the opening note and how it had lifted me away from others—that's all gone now. I am taken in by the boisterous crowd. We are one body of bodies, waving, reaching out, singing, no roof to hold our voices back from God. I feel the dampness of the man beside me and the hip of a woman jostling me. I am here with them. But eventually "*Va, pensiero*" ends. The opera ends. The actors take a bow and then the stage goes dark.

I ride alone in an empty bus, walk the streets raw as a wound, until I am back in a hotel room, a separate self once again, hard and solitary.

4

Circle People from the Moon

Four months after I met Andrew, I threw up in his car. This didn't concern me. Chronic pain often left me so exhausted that I got sick. I'd never had a regular period, so skipping several months was more normal than not. I threw up every day for the next three months, but also I was gaining weight. I blamed this on Andrew, who kept bringing over Neapolitan ice cream bars, my favorite, and leaving them in my freezer. I kept getting sick. I was tired all the time. This was normal.

People sometimes ask me, *How could you not have known?* But how could I have known? I'd been told my whole life I couldn't get pregnant. The brain takes the facts it is given and from them forms reality. I believed it to be true and so it was definitively true. Until it wasn't. At seventeen, when I had my first boyfriend, my mother took me to have some tests run so that we could be certain. The doctor explained that my disability rendered my body inhospitable. My mother had objected to his choice of the word "inhospitable" and the doctor had said, *OK, so how about "incompatible for growing a life"?*

I peed on a stick and Andrew set a timer. We danced in the kitchen, laughing, letting the timer run down. We sang in silly voices, *Our lives are over, our lives are over, one more minute of freedom, thirty more seconds of happiness, ten, nine, eight*, we sang the seconds, never thinking for any one of them that the test would come back positive. Time ran out.

I went into the kitchen. I ripped an ice cream sandwich from its wrapper. Andrew did not follow me or try to talk to me. He stayed behind, in the other room, holding the pregnancy test. I began to eat the ice cream. I could only think, *But I'm incompatible for growing a life.* I kept eating. We had no money. Andrew had dropped out of college and was driving a city bus. I was a grad student. Our combined bank accounts totaled $257. My cat strolled into the kitchen. I looked at her and thought that if our fates were reversed, she'd just disappear under the porch and push her babies out without making such a fuss. I'd like to say I then had bigger thoughts about how people in the world wanted babies and couldn't have them, people had babies and lost them, but mostly I focused on my cat and her self-sufficiency. I said to that cat, mouth half-full of ice cream, "I'm smarter than you—if you can do this, so can I." She yawned and licked her lips, unconvinced. I started to panic.

Two years before I met Andrew, my father sent an email with the subject line, *my suicide note.* It was addressed to me, my mother, the woman he left her for, the woman he left that woman for, and two more women. He told us that we—all of us, individually and collectively—had *taken his beautiful heart and broken it.*

I was staying at my best friend Kate's house. It was three in the morning. We'd just come home from a party. I showed Kate the email and she held me for a long time, then went to her kitchen and made coffee. I sat on her bed. I had an old phone number for my father and I called it. Kate sat just beyond the bedroom door. *I'm here*, she said, *right here when you need me.*

My father picked up the phone. He was drunk. He cried and said he wanted to die but wouldn't kill himself. He'd had another affair and it had cost him another marriage, his fourth, and his job. He'd been teaching at a college in Vermont. The affair had been with a student, who filed a complaint against him with the school. He'd been unwilling to leave her alone after their breakup.

He wanted me to know that leaving me and my mother was his biggest regret. She was the love of his life, he said, and I was his best friend. I believed him.

My parents had hoped to stay abroad and raise me in Nepal or Thailand or Japan or Australia. Many of the friends they'd made while traveling planned to do the same. Children were no reason to break up the party. But the reality of my disability forced my parents to make new choices. The grand adventure my father had started with my mother had ended with him in a suit, in a cubicle, in Kansas, the father to a daughter who needed more care than he knew how to give.

The day after I was born, a physical therapist came to the hospital in Bangkok and showed my parents how to gently stretch and massage my folded body. This therapist's rough touch made me cry and my father hated her. My mother did as the therapist instructed, stretching me, trying to unfurl me, and it made me cry again and my father felt angry at my mother and yet needed her so much. The therapist explained that my pain now meant less later, but that wasn't something my father could accept at that moment. He just needed me to stop crying. I did not stop. I was this unknowable creature screaming in pain.

Underneath fear and his instinct to escape, my father experienced, he told me, a rush of love for my mother that made him feel ill, lost to the depths of a new knowledge: that love and fear got all tangled up, and I thought of my father when I heard Wolfgang scream into existence, into my arms, his body wet and heavy, as I lay on the surgeon's table, my center cut open, and I listened for the first time to the sound of my son, this destroyer, and my world was undone by his voice.

My father tried to stay. He buried his discomfort in books. He read a book a day. The real world with its hard facts slipped away, but never entirely, never enough. Drinking helped. Other women helped, too. Weeks after I was born, he left my mother for a young lady in Japan with whom he chose to start a new life, a lighter adventure.

My mother kept living in Kathmandu. She laid my infant body on the concrete ledge of a fountain near our house and massaged me with oils all day in the sun. For hours and hours, she kneaded my shriveled limbs, letting the sun thaw my awful muscles.

In Japan, my father sat in a restaurant and watched a man put a live eel into a bowl of boiling water. In the center of the bowl was a block of tofu. To escape the heat of the water, the eel burrowed into the block. The man lifted the tofu with the living eel inside, sliced it up, and served it to my father.

My mother took me back by herself to the United States where there was an orthopedic surgeon in Kansas City, Dr. Asher, who specialized in spinal abnormalities. I needed surgeries on my legs and clubbed feet. The painter Frida Kahlo made a portrait for her physician, Dr. Eloesser, to thank him. Van Gogh made *Portrait of Dr. Gachet* as a gesture of gratitude. My mother took a thin strip of metal and shaped it to look like my foot. With it, she made foot-shaped sugar cookies. She painted each toenail with a thin frosting. She brought the cookies, her little portraits of me, to my nurses and doctors.

"I also gave them some cranes," says my mom on the phone. I've called her again in the evening from Rome. I call her every evening.

"What kind of cranes?"

"We folded a thousand origami cranes before your surgery."

"You and Dad?"

"No, me and your aunt Georgeanne. You know about the thousand cranes, don't you? It's a Japanese legend. Anyone who folds a thousand cranes gets to ask the gods to grant you one wish. I read you that book when you were little, *Sadako and the Thousand Paper Cranes*, about the girl who gets leukemia from radiation from the atomic bombs in Hiroshima? Anyway, Sadako wants to live and so starts folding a thousand cranes so that she can make her wish."

"And then she lives?"

"She makes half of the cranes and gets bored and gives up and then she dies. Let that be a lesson to finish what you start before you give up your degree."

"I'm not giving up my degree."

"Your aunt Georgeanne and I made all thousand cranes and we did get bored. It's a lot of cranes. One thousand cranes is so many cranes. But we didn't quit, we made them all, and we wished for your surgeries to go well and for you to be OK and you are."

"You feel like I'm OK?"

"Yeah, you're fine."

"I don't feel fine."

"Aren't you in Rome? What are you doing with your time? You know they have museums in Rome? Art, nice art. That doesn't interest you?"

I tell her about the Bernini sculpture. I tell her my memories of my father and the red-haired woman in the white dress.

"Yikes," she says flatly.

"Do you remember anything about that?"

"Was she your swim instructor?"

"No, she was a stranger in a department store. Who was the swim instructor?"

"Oh, I can't remember her name now."

"Why are you bringing up a swim instructor?"

"Robin. Ronda. Rhoda. I can't remember."

"Who are you talking about?"

"I should have clued in to what was going on as soon as he started acting so eager to take you to swim lessons. At first, he was oh so bothered to do it, and then suddenly there was no person alive more dedicated to aquatic education."

"What happened?"

"He moved out for a while and lived with your swim instructor and then eventually he came back."

"She dumped him?"

"Yeah, poor girl. She'd fallen for a myth—a kindhearted dad completely devoted to his disabled daughter. He helped her believe it as long as he could, but eventually she figured out the act. You'd played your role and now you were nowhere to be found. She didn't like that."

I tell my mother about the opera and then about the stranger in the Galleria, Joel, and how we'd followed each other through the halls.

"You were hoping he'd hit on you?"

"I guess."

"What about Andrew?"

"I didn't do anything, I just imagined it."

"Are you going to finish your degree or is that just done?"

"I am working on it."

"Is this what working looks like?"

I think to tell her about the night in the bar with Colin and Jay and what they'd said and how that night had led me here, to Rome. But I don't know enough about what I'm doing. It's murky to me and will be opaque to her.

"Dad told me that when I was born, he looked at you and knew he'd never feel a love like what he felt for you in that moment."

"That might be true," she says and then she starts laughing. "But then the next day he came back to the hospital with lipstick on his neck."

"Lipstick?"

"He forgot to tell you that story? He came back to the hospital with lipstick on his neck. You were a day old." My mother is laughing so hard I can barely hear her. "I said to him, *Yeah, you've got lipstick on your neck*."

"What did he say?"

"He said he'd gone to eat and was having a sandwich and the place just suddenly turned into a strip club."

Now I'm laughing, too.

"Were you mad?"

"He wasn't evil, he wasn't a monster, he had some limitations. He saw the world through the lens of himself. He loved me, he loved you, and I'm certain he loves you now, but his love for anyone never outweighed his concern for himself. He wasn't a person of pure intention, but neither are you. No one is. I'm grateful for what he taught you. He finished his PhD. Left his family but finished his PhD. What's your plan?"

"Mom."

"Let's be fair to him. Let's tell the truth and not a story. Everyone was part of his tableau. We all had a part to play in the story he told himself about himself. That wasn't personal to you. But, it's true, he needed you to play a specific role. You were the visible evidence that he was a good man. Because he had this—"

"Because he had this little crippled daughter—"

"And he loved her."

After my many surgeries, we stayed in Kansas, close to my mom's family, and close to the orthopedic surgeon who understood my disability. My mom got a job teaching third grade in Tonganoxie, a small, rural farm town not far from Kansas City. Eventually my father returned from Asia, wanting a life with us. He'd come and go for the next decade, then he'd leave us for good. I've asked my mother many times why she took him back and she always says the same thing: *I thought you needed a father.*

My mother never made much money teaching, but her eye and effort doubled our wealth. She embroidered flowers on my clothes, painted a sunset on my dresser, carved animals into bookshelves and the headboard of my bedframe. She was an impatient and creative cook who glanced at recipes, then riffed, making them her own. She couldn't make the same dinner twice. She wrapped every present she gave with care. Her steady hand ran scissor blade on ribbon, curling it. She layered our lives in the Kansas farmhouse with light and color. She gave me my first working theory of art—that it was personal, that it could spring from junk, that it could be present in the dull, quotidian corners of a life. She was and is a gifted painter. Her exclusive subject was horses. This stayed true until Wolfgang was born. Then she'd paint Wolfgang sometimes, but still mostly horses. She framed her best paintings but never hung them in a place of prominence in our home. One was on a bathroom wall and another sat on a windowsill in our basement. Her work was visible but barely.

My father set up a room in our house that was his and when the door was closed, no one was to disturb him. He wrote songs on his guitar, typed stories on our one computer, and read. His room was lined floor to ceiling with books. In the mornings, he took his tea and drank it in that room alone. In the afternoons, when he came home from work, he immediately retreated into that room. The walls were white and his bookshelves were gray.

He also gave me an early theory of art—that it came from the act of wresting one's singular genius out from the self and into a form made apparent to others. But he was never happy with what came out of himself, it never matched the ideal form of the thing in his mind. He wrote and rewrote the same songs over and over. He told me his plans for the many books he'd write. He started some of them, finished none. He liked the beginning of things best. The beginning was a fever. He liked possibilities. And so do I.

"What about the dog story?"

"Oh," says my mother.

"What's your version?"

"He told you his side, I suppose."

"He said you couldn't stay with him while he drowned the dog."

"That's right. That's true. It wasn't something I could see."

"He said he felt alone and that it had been a horrible feeling to be left alone to do that."

"I should be grateful to him for it. I thought it was the right thing to do, to end that suffering. I couldn't do it and he could. That's fair. I'll fess up. That was a weakness in me. I have my limits, and if he felt abandoned, well, I'm sorry."

We are quiet for a minute.

"Isn't it funny?" says my mother. "He stayed for the dog. I stayed for you."

"Why could he do one thing and not the other?"

"Who knows?"

"You must know?"

"I think he was always looking for a big feeling but didn't recognize them when they came."

"And you did?"

"Let's not bash the man. He was a good father to you when he wanted to be a father," my mother says, blowing her nose. "And he showed you what a life of the mind might look like. You owe that part of yourself to him."

My father made a classroom of our kitchen. I was his student, an audience of one, held captive by hunger. As a child, I sat rapt at the table, absorbing his lectures on the great philosophers, whose lives and ideas my father could twist up into the shape of an adventure story. He cast the philosophers as flawed and tragic heroes engaged on endless quests to conquer the dimness of others, who in turn ridiculed, imprisoned, and killed them. That these great men, misunderstood in their time, often met awful ends appealed to my father. The common man shielded his eyes from the sun; he had no endurance in the presence of greatness and so resorted to cruelty and ignorance; he took beautiful minds and broke them. My father considered this fate while filling his wineglass. He'd hold the glass aloft, lifting it into the glare from a fluorescent bulb.

"Socrates drank the hemlock himself," he told me before tipping back his glass, drinking it clean. "This was his way to immortality. Let them see they'd destroyed a good man. Let them see all they've lost and then leave them to live with that."

My father turned the history of philosophy into a story of longing. This is not hard to do. Philosophy often argues that the realest things are elsewhere. Plato's theory of forms can be framed as an articulation of pure and impossible appetite. Plato tells us there is a realm of Ideals accessible via reason alone. This realm contains what Plato wants most: perfect, ideal, beautiful, eternal Truth. But he's trapped in his present, fixed in the physical world, which contains nothing but imperfect reflections of the Ideal—mere copies of copies. The truth eludes him; he knows it exists, but he cannot grasp it. Plato is kept from the Ideal by the fact that he's a human—alive in a body, separate from perfection, suffering in actuality. Told this way, Plato takes on the élan of an unrequited lover, possessed by the desire to close an unclosing distance.

My father liked Plato's *Symposium* best, which tells the story of Socrates at a drinking party with his buddies. One by one, the partiers are all called upon to stand and give a speech in praise of Eros, the god of longing. This longing, this lust, this love that Eros reigns over—it is aroused by beauty. At first, none of the speakers can say what Eros is, only what each wants Eros to be. One wants to believe that the purpose of desire was to lead men to noble acts; another wants to believe love makes the lover divine; another claims Eros is to blame for his interest in young bodies, as the god "does not settle on what is fading and has passed its bloom."

Aristophanes tells the story of how humans were robbed by a jealous Zeus of their ancient nature. He says once we were the circle people with four arms and four legs and two faces on either side of a single head. The male of this origin story was the offspring of the sun; the female birthed from the earth. And there was a third set, both male and female. They were born of the moon. We were very powerful, so Zeus threw his lightning bolts at us, cutting us in half. This is the origin of the feeling, says Aristophanes, of missing your other half. I thought of this when Wolfgang was cut from me.

"Love," Aristophanes tells us, "is the name for the desire and the pursuit of the whole."

When it is Socrates's turn, he tells his audience of the quest for perfect beauty. The quest begins when we are young. We seek beauty, but our understanding of its nature is limited. We find it primarily in easy-to-appreciate human forms. As we grow older and learn more, we journey closer to the truth of beauty. We begin to perceive it more powerfully in minds than in bodies. We stay on our quest, ascending, going higher and higher in our conception of beauty. As we do, our capacity to recognize beauty grows larger. We can take in more. Our eyes adjust to the bright light of the true nature of beauty until, at last, we may be able to behold it—perfect beauty, which is "pure, clean, unmixed, and not infected with human flesh, colors, or mortality." Glimpsing it at last, we become part of a bigger sum, something vast and immortal.

My father was attuned to these messages. Philosophy idealized the notion of the incomplete quest, and so my father had a scaffold upon which he could elevate his restlessness. The life of the mind required constantly moving away from where he was.

Great works of art are often surrounded by intentional blankness. A museum's buffed floors, hidden switches, lights angled just so, the clean gleam of a smooth white wall—these generous and modest refinements clear the way for the grander beauty to shine. This is sometimes the worth of women to men. For some, there could never be two brawling masterworks side by side. My father was interested in the feeling he had when in proximity to beauty, and my mother's beauty suited him; it shone a spotlight on him, illuminating his worth. The narrative cohered: A great man marries a beautiful woman. She fit a vision.

This was not the sum of my father's esteem for my mother, but it was part of it. He gave this away in domestic spaces, making speeches while holding up two socks, intending, then forgetting, to fold and put them in a drawer. The laundry simply ceased to exist, its materiality incinerated in the blaze of the new idea that filled his visual field. He philosophized in life's aisles, crumpled grocery list in his fist. He'd make an attempt to locate this or that pasta sauce but would become distracted by a song writing itself in his mind. The drive home became an adventure; a missed turn filled him with a sort of delight. He'd roll the windows down and we'd drive, arriving home late and with all the wrong food, the list lost long ago.

The laundry would get folded and put away offscreen; the hands that poured my cereal and then cleaned the bowl were blurred, invisible; the grocery list, who made the grocery list? She who was rarely in frame. Always visible: my father's wild and powerful mind, his blue eyes sparking with eternal light, my father's voice, leaping from quote to joke to song. Thoughts perched themselves precariously on the edge of his consciousness and they required rescue, which could come only by the immediate reading of a passage from a book hidden somewhere in his private library. If permitted, I'd follow him and sit on the floor while he read aloud. This, it was clear, was the real and serious work.

"What do you think Dad really wanted?" I ask my mother.

"To feel validated," she says. She yawns.

"What did you really want?"

"I wanted you to have a good childhood and then, when the time was right, I wanted you to leave it behind."

"What do you want for yourself?"

"To go to bed now. Good night. I love you."

I sat in our kitchen, holding the positive pregnancy test in my hands, and I counted back the days, weeks, months from when I'd thrown up in Andrew's car and I knew right away that it was too late, that I would not legally be able to have an abortion, that I had no choice, that my choices had been taken from me, and Andrew came and stood near me, silent and pale, and I cried until I was dizzy and then I opened my eyes and tried to look into the room, but I couldn't see beyond my hands, beyond them there was nothing, only the empty space where my future had once been and I felt only the sensation of time moving, seconds ticking on into the void.

And then the kitchen disappeared.

Andrew disappeared.

The singing of birds outside stopped. Street noise ceased. Time slowed, and I heard a voice I recognized but could not place. It said, *Well. What are you going to do now?*

With the voice came a feeling of calm. It was time. No choice left but to leave childhood behind.

I looked up and Andrew was standing above me.

"I think we can do this," I said, and the color came back into his face as if he believed this were possible, that we, near strangers, with no money, could take this on, and he nodded, and we held each other for a long time. I was five months pregnant.

My mother's beauty brought her other people's attention, projected desires, and jealousies. Her Asian features prompted a fetishizing gaze. She was sometimes treated like a delicate object, docile and meek. It didn't help that she was quiet by nature. She didn't like talk for the sake of talking and she saw through people who hid their inaction behind it. I learned to arm myself against the eyes of others by watching how she armed herself. She downplayed her beauty. She never wore makeup, cut her thick hair short, and wore oversized farmers clothing that stank of horses. She was focused and efficient in all she did, working with force and urgency, pushing past people she didn't need like so much debris. If she was going to be mistaken for an object, better to be a bulldozer—solid, impermeable, powerful on the move. She saw her authentic self residing in the concrete realm of tasks; the walls of her neutral room, papered with lists of chores.

She did not show off or brag except once. Just this one time.

Andrew and Wolfgang and I were in from New York, visiting Kansas over a holiday. We walked with my mother out to the pasture and watched her horse Echo rip and gnash up patches of clover. She focused her eyes on Echo's neck and he raised his head, alert. Then she turned her gaze into his unguarded flank, and he moved until he was facing a different direction. He looked at her. She could control him completely with her stare. We were impressed.

"Wait," she said. "Watch this."

Again, she looked at Echo's neck and he turned his head. She waved two fingers in the air and said, "Side, side," and Echo crossed his hoofs and glided sidewise away from her. "Come on back," she said, and he moved sidewise back to her. "Take a bow," she said, and he politely extended one front leg and dipped his head low. She was a horse hypnotist! She was pure magic. We clapped and oohed and hooted and a glimmer of a feeling flickered across my mother's face and she almost smiled but caught herself. She went red, and she turned her back on us, but it was too late, we'd already seen her.

5

Palm Trees Revisited

The night before I left for Italy, I had dinner in Brooklyn with a man whom I'd been skeptical of for years. We had friends in common and so I often saw him at parties and group nights out. Each time I left unsure if I liked him. He was indifferent to me, which was not unusual and did not offend me. He didn't owe me his attention, nor had I expected we'd get along just because we knew the same people.

The quality of his indifference intrigued me. In conversation, he would answer my questions tersely and then walk away, offering no guise of reciprocal politesse. Once, at a dinner hosted in a friend's Williamsburg loft, I made a game of seeing how many questions I could ask the indifferent man before he'd ask me one in return. It was a one-sided game played for my secret amusement, making it an ungenerous game. But I was feeling ungenerous. I asked the man where he grew up, did he have siblings, where had he traveled, did he like living in Brooklyn. I got to thirty-six questions, all of which he answered while looking mostly at his phone. As I started in on the thirty-seventh, he received a text about another party and he left.

The host of this dinner told me later that the indifferent man had long felt that having children was ethically reprehensible and had a vasectomy at nineteen to ensure he'd never procreate. The host told me this to let me know that, perhaps, the indifferent man was not indifferent to me at all, but in fact resented having to sit next to me, a mother.

I found his disdain exciting. He projected a radical social freedom

I felt I could not access for myself. The man did not soften the edges of his actions and felt free to wield his principles like a weapon and, although I was the recipient of the blow, I welcomed it as instructional. I'd be further in life if I'd claimed more freedom and had not so fully absorbed the lesson that my social value depended on my being compromising, deferential, small.

The man's cruelty brought comfort; it shored up the story I told myself about myself and validated my deepest suspicion: I was invisible. The indifferent man gifted me the satisfaction of being right.

I invited the indifferent man to have dinner with me, certain he'd say no—and that he'd say it just like that: *No!* But to my surprise, he agreed. We met at a Chinese restaurant. At the end of the meal, the subject of a mutual friend and his new girlfriend came up. I'd been told by others that this mutual friend's new girlfriend was the most beautiful woman in Brooklyn. I said this to the indifferent man and he shrugged and showed me pictures of the woman, our mutual friend's new girlfriend, on his phone and she was, as advertised, quite beautiful. I was struck by how accurately I'd imagined her without any prior, concrete description. Her body fell within a familiar range. She was tall, but not so tall she might emasculate, and thin, but with soft weight in the right places. I found myself wishing I could touch this woman's hair. It was long, auburn, and had a wild quality to it. The unruliness of her hair granted her a specificity so that she did not just evoke the general concept of beauty, but rather a single instance of it. She was young, white, symmetrical, able-bodied. A million times I'd seen this body shape and variations on this same face and been told, and often agreed, that it was beautiful.

Looking at the picture on the indifferent man's phone, I'd wondered, not for the first time, what my life would have been like had I been born with this woman's hair and face and body. My recurrent thought—seductive and unsound—was that I could have had anything I wanted.

I asked the indifferent man why he had pictures of our mutual friend's new girlfriend. He said he'd dated her too, just a few months before she got together with our friend.

"How long did you date her?" I asked.

He shrugged. "A few months."

"And what happened?"

"I didn't find her attractive," he said.

"But don't you think she's beautiful?"

He shrugged again, then told me stories about being at bars with other male friends who were able to pick out five or seven or ten women who were attractive to them, none of whom he, the indifferent man, would even consider.

"A 'ten' in the eyes of most men will be a 'six' to me. The girls our friends date are all, like, a 'three' for me," he said. I sensed a reluctant pride. He needed a really beautiful woman, a supernaturally beautiful woman, he explained. He leaned in to make me his conspirator.

"This may be more than you want to know," he said, "but if a woman is not, like, model-beautiful, I can't even keep up an erection when I'm with her."

A few feelings in me collided. I was disgusted by what he was saying, but I wanted him to keep talking. My academic training told me to attend this confession as it teased the difference between defining beauty and defining what beauty does in the body. The latter question belongs to the realm of aesthetics, the study of bodies in proximity to beauty. Awe, arousal, exhilaration, tumescence or the failure thereof. Aesthetics, from the Greek *aisthetikos*: sense, sense perception, sentience.

It was clear he could confess to me the details of these sexual-aesthetic experiences because I was not on the same plane as these other women—the threes, the sixes, the tens. My body barred even my consideration, and I was absent from the realm of possible women. Seeing this brought with it orienting relief, as if I'd been looking in a distorted mirror and someone had just replaced it with a normal one. What he reflected back wasn't kind, but it was clear. The indifferent man offered no excuses or apologies.

"It's a curse," he continued. "I'd like to be able to date more women. But you cannot control such things."

"Can't you?" I said.

"Of course not," he said.

I thought to leave the restaurant. Instead, I said, "What you are describing is so superficial."

"Yes, it is," he said. "I know where it comes from. I'm controlled by media and advertising." He rolled his eyes. "*Et cetera, et cetera*," he added with a trill of laughter. "But I can't be at fault for being a product of my environment."

"But if you are aware of these external, negative forces influencing your behavior, you could think about them and learn to adjust your behavior."

"How?"

He asked this sincerely. A fluorescent light flashed outside. Did I really believe, the man was asking me, that we could use our intellect to unlearn our cultural training? To unsee what we've been taught to see? I felt unsure. His gaze was convincing. I wondered why he'd agreed to this dinner. I was likely the least physically attractive woman whose full name he knew. Were we here to make sure I knew exactly that?

"Does experience eventually alter the immediate effect of beauty?" I asked.

"No," he said.

"But haven't you met someone," I asked him, my voice rising, "who seemed very beautiful at first, but who was, I don't know, boring or not very kind, and so you lost interest? Haven't you gained interest in a woman you'd, at first, too quickly overlooked?"

He shook his head. He'd grown tired of the conversation.

"You can't undo what we're taught from birth," he said. "Beauty is what we're told is beautiful and what we're told becomes the truth."

I am thinking of the indifferent man on my train to Lake Como. I'd left Rome for Milan but I had a few days before *the experience* I'd come for, and so I decided to go to Lake Como in the interim. I was being led by instinct, but also by an ad I saw while scrolling through social media. The ad told me Lake Como was the most beautiful place on earth. I closed one app and opened another and booked a hotel and train ticket, and now, through the train's window, I see Milan ease out of sight like a heavy mark being slowly erased.

I watch the landscape break apart. The terrain gradually becomes rockier as if the mountains need time and space to gather themselves up before climbing out of the fields and rising above the rooftops of the fading city.

A boy sits across the aisle from me. He turns his back on the window and tries to sleep, ignoring all unfolding behind him. A few times, he blinks himself awake and looks inward, into the train, down the aisle. He knows what I don't, that the train is minutes from Varenna Esino and will stop there only briefly and so he's made plans for the open aisle ahead, visualizing his body up, out of his seat, through and off the train. A woman near me trades her heels for a pair of sneakers, then takes out a book and reads.

The mountains now are high and close. Fog burns off at the crests and vapors join the pink underbellies of passing clouds. Sunbeams creep down the cliffsides, sending silver fires flashing through the trees. Foliage shivers and foams. Leaves, courting the wind, wink, shimmy, take breath. Then, without warning, the lake is upon us. I notice it first as a change in the light. No longer constrained by mountains, the lowering sun breaks freely on the water, drawing sharp outlines along lapping waves. The faces of strangers on the train look new. I see now a faint scar at the boy's temple and the woman reading the book is younger than I'd first thought. Her mouth appears bisected, half in shade, half illuminated. Her lipstick is red in shadow, magenta in the train car's new brightness.

My stomach wheels. I stare out the window and feel incredulous, almost angry at being swept up and carried away by the force of this beauty. I remember someone once told me that every painting was a solution to the problem of how to best carve light, but it is only at that moment on the train that I begin to understand what that could mean. This beauty brings with it the illusion that greater meaning is close and real, a grape I can pluck from the vine.

My way of looking is changed, my perceptual discernment is refined. Every painting from now on would bring with it, if gently, the memory of being on this train watching the lake spread out before me, the light changing all I'd just observed. I see dust particles dancing, a static laid upon the scene. The passengers glow as if lit from within. I see their bodies plainly, the curves under their clothes, where their skin is smooth, their sweat, the whiteness of their eyes. The woman with the brilliant lipstick glances up from her book, tilts her chin to the window. The boy with the scar no longer looks so bored. He watches the lake with longing, as if he wishes to strip and dive in. The sun penetrates our train car and I feel it, a heat on my body, and I know that the passengers can now see more of me, too. The sound of the wind through the trees mimics the sound of waves or the waves mimic the wind through the trees. The train ducks behind a grove and shadowed patterns of leaves grow across our laps. I hold my breath. I've never seen light carved so beautifully.

Off the train, down a short hill, and I'm at the edge of the lake. I stand there for a long time, my suitcase at my feet. I watch the trees and the glittering lake beyond the trees and the ring of mountains beyond the lake.

My hotel is a twenty-minute walk from the station. I pass through the town square and see people outside, drinking and eating and watching musicians play on the steps of a medieval church. A group of older men sit outside a café, smoking and clapping along to the music. Pink light refracts through their wineglasses, watercoloring their white shirts. They lean toward each other, laughing furtively behind their hands. The air is oily with the scent of rosemary that grows wild in big bushes all around the lake. If only someone would beckon to me, hand me a glass and gesture to a chair, I could abandon my walk, my suitcase, and join them.

My right hip aches, as it has with more regularity in the past few years, that pain worsening. As I walk through the square and up a hill toward my hotel, the ache focuses until it becomes like a thin blade wedged into my hip joint. The muscles in my back and along my spine bear what feels like a thousand small rips.

The driveway up to the hotel is steep. I inch up sidewise, one foot shuffling beside the other. Halfway up, I lose my grip on my suitcase and it rolls on its wheels a few feet before toppling over and skidding to a stop. I am glad no one was around to see me, their gaze relaying back to me my perceived helplessness. But I'm not helpless, I'm struggling. People don't always recognize the difference.

My hotel is a former monastery built into the side of a steep mountain. To reach my room requires a ride in two rickety funiculars. It is dark now. The lake has disappeared, but I can see the twinkling lights of Varenna below. I open my windows and voices drift in. I want to be with people, but pain commands the moment. The sound of laughter echoes in the valley. I stand above, separated from it. For so long, I'd told myself not to travel—better to stay home than to venture out and seek something that would only exclude me.

I think about the indifferent man and how he'd spoken to me. Sitting across from each other in the restaurant, he'd held the posture of someone waiting for a timer to expire. He probably didn't enjoy being seen out with me, lest someone mistake us for a couple. Or perhaps that had been the motivation behind agreeing to dinner. Male friends told me I was an excellent prop. Their close association with me, some sad-seeming cripple, made them seem sensitive, good-hearted. A past girlfriend of a man I knew once pulled me aside at a party to tell me that my friendship with her boyfriend was what had drawn her to him in the first place.

She'd said, "It was nice to know that he could care about any kind of woman."

The indifferent man had spoken about beauty and ugliness in objective terms. The absence of beauty took objective, measurable effect, robbed him of his erection; whereas the presence of beauty inspired in him love and devotion. He'd said all this calmly and without judgment. He smiled to let me know his honesty might be helpful to me. What a pity for me to labor under false hopes.

He was neither the first nor the last man to approach me in this paternalistic mode. Men did not like me getting the wrong idea. There had been a boy, Jim, in high school who had taken the time to inform me of my worth early. He and I were in a close circle of high school friends. We dyed our hair pink or purple or orange, had face piercings and handmade stick-and-poke tattoos. I have a permanent reminder of this era: a wobbly drawing of the character Harold from Hal Ashby's film *Harold and Maude* that my friend J.B., after a few beers, did freehand on my forearm with a razor blade dipped in India ink. In the film, Harold—young, disillusioned, misunderstood—falls in love with eighty-year-old Maude. Their romance is strange, specific, joyful, socially unsanctioned. I identified with both characters. Like Harold, I, too, felt disillusioned and misunderstood. Like Maude, whom I tattooed on J.B.'s arm, I was regarded as a socially inappropriate object of romantic desire.

Romance was for me an incongruent concept. I was a teenager and I wanted romance and I wanted sex, but I was often taken as disqualified from both. I would listen to other kids my age talk about attraction in terms of a sliding scale—someone could be sexy or not or sort of sexy or wasn't sexy before but was kinda getting sexier all of a sudden. My disability kept me, in the eyes of others, off that scale altogether, like an animal or a child.

But Jim knew me well and cared for me, it seemed. The shortest boy in our friend group, he liked to stand near me because it made him feel tall. He was an excellent dancer and taught me moves. I was the one he could lift the highest and with the greatest ease, swinging me through his legs, flipping me around his arm. Afterward, he'd hug me and kiss my cheek.

Our school held a homecoming dance and everyone in our friend circle began to pair up. I didn't have a date and Jim didn't either. It seemed obvious, for purposes of group unity, that Jim and I would agree to go together. But it was also well-known that I had a crush on Jim. When the subject came up, Jim would look at me expectantly. My girlfriends would ask me at the end of each day, *Has he asked you yet? Has he asked you?* And each day, I'd say, *Not yet*.

At night, I'd imagine the moment when Jim would lift me in the air and spin me around and kiss my cheek and ask me to be his date. All the angst I'd curated as a teen dissolved into romantic reverie. I, like everyone else, wanted to be chosen. As the dance grew closer, I became anxious. Our friends had started to make plans for the dance and I wasn't a part of them. Finally, my girlfriends urged me to stop waiting and just ask Jim myself.

I approached him in the library of our school. He was studying for a geometry test. He saw me, closed his notebook, and smiled.

"I feel like," he said, teasing me, "there might be something you want to talk to me about."

I told him yes, there was, and I said that I wanted to go to the homecoming dance with him and would he take me.

"Of course," he said. Relief flooded through me so quickly it turned my stomach. "But," he continued, "there's something very important I need to talk to you about first."

He proceeded to tell me that our female friends had been pressuring him for weeks to ask me to the dance, not wanting me to feel left out.

"They love you," he said, "but they pity you and their pity won't help you in the world." I can, to this day, recall the exact even tone in his voice, his smile. He reached across the table and took my hand. "I want to tell you something as your friend," he said. "I want to protect you. When you ask a man like me on a date, you put us in a bad position." He was still smiling; I was having a cute delusion and was in need of his loving, if uncomfortable, correction.

"It's just the truth," Jim said. "No man will want to date you unless he, too, is desperate or ugly."

When men I met, later in life, decided I was sexually alluring, they would often describe it as a shock, an unforeseen surprise, like they'd found an old coat with a twenty-dollar bill in the pocket. The word "actually" was commonly employed. *You're actually attractive.* A stranger had once stopped me on the street and said, *I'm not sure what you have going on here with*—he had pointed down the length of my body—*this. But I actually think you're pretty.* He asked for my number, which I declined to provide. This was not the response he'd expected. He followed me for a few steps. *No, no,* he said, diagnosing the problem, *I actually want to take you on a date. I'm actually being serious.* A man had fucked me once and after said, *Wow, you feel just like an actual woman.*

Being outside the realm of the actual came with certain freedoms. I felt no pressure to build a grand narrative around the act of losing my virginity. I didn't need to be in love and the moment did not need to be special. It could be exactly what I wanted it to be—a biology experiment, an academic exercise.

I met Patrick the summer before I left Kansas for college. He was a freshman at the nearby state university. I was invited to a small party at his dorm. When he saw me, he approached and said, "I have a whippet just for you." I excused myself to text my best friend Kate to ask her what a whippet was. Patrick took me to his room and closed the door and gave me what just looked like a balloon, and I put the opening of the balloon to my mouth and breathed in and waited but felt no change. I was certain a joke was being played on me.

He said, "I have this if you'd prefer it," and handed me a bottle of cheap whiskey. We drank from the bottle, and he pressed me against a wall and kissed me and I thought, *Ah, here it is, the college experience. I am in the midst of a classic college experience.*

When he entered me, I thought only of the strangeness of the sensation. It was neither painful nor pleasurable, or it might have been both, but those sensations were overtaken by something more important, which was the sensation of newness. I was shocked that my body, which I knew so well, could produce a wholly new feeling.

I did not see this act as real, nor did I think of Patrick as a real person during this experience. I was so far above our bodies—observing the mechanics, taking mental notes. So, when he asked, after, if I'd had an orgasm, it struck me as a funny if irrelevant question. He asked me again and I felt resentment at being expected to answer. He looked at me insistently and I performed a noncommittal nod.

"You didn't," he said, and I thought, *What business is it of yours?* I'd already retreated to the neutral room where I was having a debrief with myself. His questions were intruding on my solitary reflection on the results of this experiment. He moved my knees apart and put his tongue in me and, after a few seconds, came up and said, "There's a bit of blood here."

"Oh?" I said, and felt genuinely surprised, not because I didn't know that blood would come once the hymen was broken, but because I'd assumed I'd be the one exception, the one virgin who didn't bleed. I did not think my body would betray my mind and reveal a private piece of information.

He touched the blood on his sheets and asked me if I'd ever had sex and I shrugged. I didn't consider that this was a piece of information that might have caused him to make a different set of choices. I didn't really consider him at all. This was not an experience we were having together. We were a different species.

The next morning, I wake with the dawn and walk the edge of the lake. The beauty of Lake Como is so massive and all-consuming that it accosts me. I turn a corner and cringe. I curse. The beauty of the lake is absolute, resounding. Birds sing, waves lap, the air smells of fir and jasmine. The sun shines, but it is not too hot; the wind blows, but it is not too cold. There is something restorative, palliative, in the air. I can walk further, the twinge in my hip that I'd felt worsening for so many months is gone. I can sleep more, eat less, drink less, hear better. I take off my glasses and can see a great distance.

I'm easily seasick, but I want to see the other villages sprinkled around the lake. With fear, I buy a ticket for the ferry and stay on it for hours, taking in the grandeur of the mansions perched along the shore, the old churches, the verdant mountains. I feel not so much as a twinge of nausea. I take the boat to Bellagio where the English during the Victorian era flocked to recover from tuberculosis. To be here, to witness the lake, seems suddenly the most logical prescription for any ailment.

I sit on the shore, eating pizza and then gelato. The sun sets. I watch the palms at the edges of the water. I think of the scholar Elaine Scarry's writings on the beauty of palm trees. In her book *On Beauty and Being Just*, she argues that there are two common points of error in our perceptions of beauty. The first, which she refers to as an error of overcrediting, occurs when we recognize that something we'd formerly thought beautiful no longer is. The second comes when we realize we've withheld the attribution of beauty from an object that has rightfully deserved it all along.

"For example," she writes, "I had ruled out palm trees as objects of beauty and then one day discovered I had made a mistake." This latter error, the error of undercrediting, is the more serious, as it is evidence of our "failed generosity."

Scarry recounts a moment when she is on a balcony watching the leaves of the palm tree move—they are "lustrously in love with the air and light." Her perception shifts, a mistake is corrected. "The vividness of the palm states the acuity with which I feel the error, a kind of dread conveyed by the words 'How many?' How many other errors lie like broken plates or flowers on the floor of my mind?"

Night comes and Scarry is still on her balcony watching the palm tree, witnessing the missed beauty. "Under the moonlight, my palm tree waves and sprays needles of black, silver, and white; hundreds of shimmering lines circle and play and stay in perfect parallel."

I never much cared for the early Greek concept of measurable beauty, nor did I agree that it was a virtue on par with truth or justice. I'd always preferred Hume's notion that beauty was not a set of external properties, but rather that it existed in the contemplative mind.

"Each mind perceives a different beauty," Hume argues. "One may even perceive deformity where another is sensible to beauty." And this is a better theory for me, a woman with a body that could never be mistaken for symmetrical or orderly.

In contemporary aesthetic scholarship, the Greeks' objective evaluations of beauty are often regarded as wrong, outdated, or—even worse—uncool. But as I lay there in Varenna listening to the sound of the lake's gentle waves coming to shore, I wonder if I'd rejected the possibility of divine, objective beauty simply because I was excluded from it. Being excluded from a theory did not make it incorrect.

Elaine Scarry wonders how she's missed the beauty of the palm and decides it is because she'd held in her mind some sort of palm tree composite, made of, maybe, television images, bad drawings, blow-up plastic palms at party stores, and that composite was ugly. It becomes beautiful only when the individual, specific elements line up, when the light is just right. It isn't palms, but *this* palm that corrects her error. It's the singular instance of one palm. She looks up under a canopy of palm leaves and sees an owl sleeping. By weaving its plumage with the palm's leaves, the owl can suspend itself midair to sleep as if in flight. *Beautiful*, she thinks. So now it is not just this palm that convinces her. It's a set of particulars. The palm, the minute, the coolness of the spot where she stood, looking up. The noticing of the owl.

The next day, I go to Villa Monastero and walk through the great gardens there. A brochure tells me that several species of rare plants are able to flourish on the steep slopes of the botanical gardens in Varenna and among them are certain varieties of palms. African and American palms, Chilean wine palms, European fan palm, Mexican blue. With Elaine Scarry in mind, I find myself very attuned to the beauty of these trees, especially the Mexican blue palm, whose blue-gray fronds shimmer and stir, thin fringe waving in movie-light. The whole scene is beautiful to the point of looking manufactured. When I look out over the water to the painted mountains, I feel certain the scene will flicker and reveal a green screen. The brochure explains that the Mexican blue palm is somewhat rare, able to flourish around Lake Como due to the perfect climate, mild year-round.

I hike into the mountains. The path I take is so steep that at times I get down on all fours and use the rocks to drag myself up, like a mountain goat. At the summit is yet another monastery. I climb its turrets and look out at the paradise below. The sun stitches sequins on ridges of waves. This is what the poet Longfellow had looked at when he'd written, "I ask myself, Is this a dream? / Will it vanish into air? / Is there a land of such supreme / And perfect beauty anywhere?"

Beyond the monastery I find myself surrounded by trees. Ahead, two older women, grocery bags in hand, are cursing in Italian at a bush. I hear a rustling. A third, much deeper voice emerges, laughing, and the rustling speeds up. The women curse louder, and the deep voice laughs louder. I approach the scene. There, in the bushes, a man is masturbating. His pants are around his ankles and his thin penis sits gently in his palm. It has a gray and waterlogged look, like it's been submerged in the lake for days. He makes kissing sounds at the women, but when I pass into view he freezes. He looks at me for a moment then pulls up his pants. He glances down at my shriveled legs, visible below the hem of my dress.

"Mi dispiace," he says. *I'm sorry. "Dio si benedica, Signora." May God bless you.*

Longfellow at Como: "But ye have left / Your beauty with me, a serene accord / Of forms and colors, passive, yet endowed / In their submissiveness with power as sweet / And gracious, almost might I dare to say, / As virtue is, or goodness."

A few months after the homecoming dance, Jim tried to kiss me.

He said, "Something has changed. And I like you."

Before, Jim had spoken to me as if beauty were an unalterable fact, but now suddenly my beauty was discernible, visible if only in the right light.

"What changed and how?" I asked him.

"You grew on me, you made me laugh enough times that I started to want to be around you more, you are smarter than my last girl."

I remember the pang of pride I had when Jim said this. I remember how it motivated me, like a dog wanting to please its owner, to prove my worth to him over and over again.

Jim's perceptual shift, not what he said in the library, embedded a damaging idea in me, one I'd recognize deeply when I read Scarry years later: beauty was a matter of particulars aligning correctly. My body put me in a bracketed, undercredited sense of beauty. But if I could get the particulars lined up just right, I could be re-seen, discovered like the palm tree is discovered. To be deserving of the whole range of human desires, I had to be extraordinary in all other aspects.

In this new light, I started to see my work, my intellect, my skills, my moments of humor or goodness, not as valuable in themselves, but as ways of easing the impact of my ugliness. If only I could pile up enough good qualities, they could obscure my unacceptable body.

"The correction," Scarry writes, "the alteration in perception, is so palpable that it is as though the perception itself (rather than its object) lies rotting in the brain . . . the perception has undergone a radical alteration—it breaks apart or disintegrates. . . ."

Philosophically, Hume and Scarry provide richer views of beauty than the Greek conception of mathematical perfection. But accepting the argument that beauty was malleable came, for me, with a cost. The Platonian view rejected me cleanly, but Hume and Scarry left a door ajar and I've spent a lifetime trying to contort my form to see if I could pass through it.

I liked sex because I liked acquiring knowledge. I went to college in Boston where I met men willing to provide me with valuable data sets. After I graduated, I moved back to Kansas, got a job stocking shelves at a grocery store, and lived with Kate. We spent all our time together. We went out at night, made meals, but mostly we talked. If I was up before her on a weekend morning, I'd crawl into her bed and wait for her to wake so that we could start our day of talking. My interest in dating, flirting, sex, was, at this time, all in service to the realest thing in my life: talking, gossiping, laughing with Kate.

My job at the grocery store started at five a.m. I spent the morning turning soup cans so that their labels faced outward. After work, I'd go to the library and borrow DVDs. Over and over, I watched *Casablanca*, *Sabrina*, and *The African Queen*. I'd lay on our apartment floor in the late afternoon, rising in and out of sleep while Humphrey Bogart charmed someone on-screen. When Kate came home, she'd lay on the floor with me and we'd listen to records and recount our days and forecast the night to come, which, like all our nights, would be spent in one of the two bars we liked in town. We went to these bars and drank cheap beer and bummed cigarettes and listened to local bands whose members were our friends. There was an intoxicating sameness to this. In Boston, time had marched forward, but in this drowsy year, back in my Midwestern hometown, repetition cast its lulling spell and time stopped advancing. Each day and night was merely a rehearsal for the next. I was always tired, always a bit dazed; I cycled from sleepy to drunk to hungover and back. It was a blissful languor. I had no responsibilities. I put cans on shelves for a living. I made just enough money to pay my rent and to buy beers. My choices and mistakes held no value nor consequence. I felt little, thought less. I had no ideas. My aspirations stretched to the closest guy with good hair and no further.

One night, in the basement of the Taproom, Kate introduced me to a man I'd not met before. This in itself was an event. Mostly we knew everyone in the basement of the Taproom. But this man, I'd never seen him before, which was so unlikely in a town so small that he immediately took on the air of a magician. He spoke to me with immediate interest, turning away from his friends. I waited for his disappearing act. He was so handsome it made me feel a bit ill and I nursed the drink he bought me. We sat, huddled together in a corner, and then the lights flooded on, signaling the end of the night and our need to leave, and it was only then that I realized that hours had gone by without my notice.

The man told me he was a barista at the café next door and on Tuesdays he was off at noon. He asked me to come see him. That Tuesday, he took me to lunch. After, he told me to come again the following Tuesday and I did. We went to a bookstore nearby and I bought him a Raymond Chandler novel. When we parted, he took me in his arms and kissed my cheek and in my ear he whispered, "Tuesday?" And, again, I went to see him. I'd always come a minute or two early so I could watch him finish his shift. I liked the way he wiped down counters. I liked his smile—equal parts mischief and delight. We'd eat or play board games in the café or drive around in his car. He'd read me poems, show me paintings. He fingered his record collection, looking for the perfect song to play for me. For the first time, I felt I was meant to be included in a man's enthusiasm rather than merely instructed by it.

The world split into two states of existence—Tuesday and not Tuesday. My days working at the grocery store became increasingly blurry, dreamlike. Mornings fogged with the thought of him. The aisles of the grocery store become quiet cathedrals of desire. I could walk them alone, free to exalt him. My senses sought his every reminder. The store, empty and dark, took on an erotic hue. I slid the blade of my box cutter from its sheath, sliced through taut tape. The cardboard petals of the boxes unfurled, revealing color—greens, yellows, reds; clean cans of food. I'd hold a can of yams in my hand, smoothing my fingers along the metal's edge. Some cans came dented, seals broken, and their interior scent oozed into the atmosphere, made me woozy. Minutes moved, dim light brightened, people arrived to shop, but none of this diverted my focus, nothing distracted me from my thoughts of him; I was a disciplined instrument forged by wanting.

When we were together, the man stood close to me. He'd held me, hugged me, kissed my cheek chastely when we said our goodbyes. Sometimes he looked at me like he could see through my clothing, like his eyes could touch the softest parts of me. After a year of Tuesdays, I descended the stairs and stood in the basement of the Taproom and there he was, across the room, watching me. He nodded, I nodded. This moment had always been on its way and now we recognized its arrival.

We said nothing. We walked together from the bar. Snow lurked. It was winter. The pungent scent of incipient precipitation surrounded us.

My apartment was dark. He moved inside and said, "It's started." He was at the window, watching the newly falling snow. Outside, lamplight shone on snowflakes, ellipses written onto darkness. We removed our layers slowly. He took off his hat, a glove, another. He moved to his boots with hesitation, holding the laces aloft as they came untied. I saw the question in the air. To retie the laces or let them fall? He was making a decision about what he wanted to happen next. When he sat back up, he was blushing. The moment was delicate, it held us in its crystalline intimacy. Us, in our socks, alone.

We moved closer slowly. The light outside shifted not imperceptibly to me. Behind my eye came a shooting pain. I must have grimaced because he reached for me then. He smoothed my hair and held my face—lightly first, then harder. He pressed his fingers into my temples and, with his touch, the pain behind my eye was gone, but returned immediately when he released me, a thudding reminder that relief is fleeting. His hands were rough with work. There was paint beneath his fingernails. He smelled of coffee, cedar, juniper, gin. The edges of his lips turned down when he smiled, making his happiness look incredulous. I wanted him to touch me and then, reading my mind, he did. He touched my lips with his thumb, he said my name, sweat slipped from my temples, lost in the dark; I was glad for the darkness, not because it hid my face, my fear, but because it hid part of him, dulled him just enough that I could hold myself together. He was so gorgeous. To face him in full light would have unraveled me. I felt a rising feeling now, desire, and then, at desire's crest, he leaned forward to kiss me. His lips almost met mine but missed. I'd pulled away.

He blinked with surprise, his face flushed red. "Oh," he said. "I'm sorry."

"No," I said, but it was too late. He stood up. The moment fell softly to the floor.

"I didn't mean to—" he began.

"No—" I said, again, cutting him off, but couldn't say anything else.

"I misunderstood, I'm sorry, so sorry, I thought we—" he said.

"We did, we do—"

"I've got to go home now. It's late." He couldn't look at me. He went to the door as fast as he could and when I stood and walked toward him, he instinctively flattened himself against the frame as if opening the door would take too much time and he needed to phase through it.

I wiped the window to see him, out on the street, racing away. This man, handsome as a movie star, kissing me in the moonlight? I don't think so. I was no Mary Astor, I was no Bacall, there was no one like me in the movies, no educative model, no narrative that included a body like mine in a grand romance. Didn't I know I was for conflicted men only, the ugly and desperate? I did know. I'd learned my lessons well. And I was grateful for them. I was safer inside a shell not punctured by hope. I was sorry to see this man go, but there was no way around it. I had a worldview to uphold.

I met Andrew six years later. He kept appearing wherever I was, staring at me, refusing to speak. If forced, he'd answer questions with single-word responses. One night we were drunk and slept together. No snow softly fell. No strings swelled. When Kate met him, she said, "Well, he's cute, but how serious could you be about him really?" And I answered honestly that I wasn't serious about him at all and had told him that I hoped to sleep with him a few more times and then never think of him again.

Andrew received me calmly. He could create a pool of silence around himself and then dissolve into it. He could turn down the lights behind his eyes and could just be gone. He could board up his entire being. When he got like this, one look told you you'd get nowhere with him. Sleeping people had more to tell you than he did when he zipped himself all up. This tricked me into thinking nothing was happening, nothing was taking a hold of me.

He fell in love with me quickly, which I took as a sign of weakness. But then I fell in love with him quickly, too. It felt like betrayal. I knew this trick. I knew it led to my suffering. I waited for weeks for him to start laughing at me. I waited for him to say *Just kidding, you fell for that, what am I, some ugly, desperate loser?* But the weeks went on and the more time passed, the more I wanted to provoke him into saying *OK, OK, it was all a joke.* Our sex was notable in that I couldn't get outside or above it. He would touch me, and I could be nowhere but with him. He pulled my mind inside my body and made me live inside it. I didn't like this. I missed him when he wasn't with me and this made me angry. I was waiting, waiting, and while I waited, I invented his cruelties, pulling them from thin air. Once, in a department store, he stopped holding my hand to reach something from a shelf and then he didn't take my hand again and I was certain it was because he was embarrassed to be seen with me in public. And I raged at him that night, calling him every name I could think of, begging him to show me how little he cared.

He listened to me, seeing right through my words to my fears. He didn't flinch or become upset. When I was finished, he said, firmly and without judgment, "You are hurting me, and I'd like you to stop." Anger burned in my chest. He saw it. He saw me. He did not run. He stood in front of me. He was waiting, too, I realized.

I thought others had to, first, unsee my body and then, only then, could a door be unlocked and a realer me let out. But when it came to love, desire, sex—I needed my partner not to unsee but re-see my body, discover they desired it. This was too easily done by Andrew. He wanted me fully and without complication. Not only did I not trust this, but it exposed me. There's safety in being locked away in the dark. Other people's dismissal of me, their discomfort around me, their unwillingness or inability to see me as anything other than a walking tragedy—this all absolved me of a certain responsibility. I was free, in a sense, to live outside the realm of people. And I was free to feel superior to people who would exclude me. It is hard to leave the darkness and step out into bright and baring light.

I looked at Andrew and saw a man I loved and was treating badly, and I realized the era of inconsequence was over. The romantic wounds I'd nursed, my melodrama and self-pity, could harm someone. And Andrew had the grace to simply tell me this and offer me the chance to change. He let our days together unfold, not cinematically, but actually.

Church bells clang, echoing strangely out across the lake, colliding against the mountain walls, playing a mangled melody. The bells are chaotic, sharp, abrasive.

I finish my final meal in Varenna. I walk up into the mountains—I'm stiff from my hike the day before and soon my body is more than stiff, it is in agony; I walk and walk, up higher and higher. What I want is pain, what I want is to feel the ugliness of myself, my body and mind; it scratches an itch, it confirms what calls in me for confirmation.

I watch the sun disappear. I will it to set slowly so that I may linger over this perceptual moment, to somehow swallow it up and keep within me forever this view of the lake, the mountains, the flowers, the palms. The next morning, my last, I sleep too late, miss the sunrise.

As the train rattles away from Lake Como, I close my eyes and see, not my husband, not Jim, not the man from the snowy night, but the indifferent man. I see myself naked at the edge of the indifferent man's bed. I hear him tell me what he sees: that I am too ugly to fuck. It isn't my fault or his, merely a fact, impossible to undo. It is a regrettable truth, a door firmly shut. I imagine this validating cruelty, my familiar friend. It is what I really want. Not his sex, but the intimacy of his decree, not hidden in sympathy. I want to see his bare face. I imagine how beautiful it would be to hear someone speaking plainly and without subtext to me.

I've seen photographs from performance artist Marina Abramović's *Rhythm 0*. She'd stood, silent and still, for six hours next to a table with seventy-two objects laid out upon it. Feathers, flowers, wine, knives, razors, a loaded gun. There were written instructions for the audience to use the objects on her as they desired. *I am the object*, she wrote. *During this period I take full responsibility.*

The performance began tamely. People offered Abramović a tickle or a kiss. She looked at no one, reacted to nothing. A few hours in, people cut her clothes off her body with razor blades. Someone nicked her throat and drank her blood. She was groped and hit. The loaded gun was placed in her hand and positioned to point at her head. Someone else took it away. An argument broke out in the crowd about where to aim the gun. Abramović later described the fear she felt, the violation. But I wondered if she'd also felt elated to be awake and alive in the filth and debris of humanity revealing its nature. Someone had run the edge of a knife along her skin. How horrible, but also how real, how bright, that touch would have been. There was a power in being the object that pulls out people's worst selves. There's pleasure in moving away from the hopeful unknown and into cruelty's indifferent absolute.

When the six hours were over and the performance ended, Abramović started walking around, herself again, no longer an object, and everyone in the audience ran away. People could not stay and face her once they saw her as a person.

A last glimpse of Como and it's gone. Milan looms. In a week's time, I'll be back in Brooklyn. When I return, the indifferent man will ask me to dinner, and we will meet again at the Chinese restaurant where he tells me he's met a woman at a party and has fallen in love. I will meet the woman. She is beautiful, brilliant, and kind. Later, he will reverse his principles and his vasectomy and she will give birth to the indifferent man's child, a baby girl. He's different now, the indifferent man—this man who dismissed me, judged my motherhood: a father now. What changed and how? No matter. Time eroded and rewrote his convictions. Change came like a wave and sent his old self out to sea. I would like to gloat, but of course I feel only an instructive fear. I'm not safe from fate's revisions; I, too, stand on an unstable shore, waiting for the tide to come in and take from me what it will.

6

The Weakness of the Spectator

I fall asleep on the train from Como back to Milan and dream of Wolfgang. It is a recurring dream, always the same, beginning with an abrupt nothing, a quiet that is not the absence of sound but sound's inverse, a ringing emptiness that tells me something is wrong. I listen for my son—his cry or heavy breath. I'm afraid something has happened to him and afraid to have my fear confirmed. I try to move toward his bedroom door but I can never make it to the end of the hallway. The dream is relentless this way, keeping me stuck, suspended in gray light, moving but not forward.

I wake to the thud and rustle of suitcases. The train has arrived at Milano Centrale. I transfer to a bus that will take me to my hotel. Milan looks like Midtown Manhattan to me, gray and goal oriented, a city full of people on their way to work. I watch a woman power-walk across the street and feel my own confidence grow by observing her sense of direction. She looks just like my mother, not her face, but the way she moves.

I check into my hotel and lie down on the rough, utilitarian carpet. I'm stiff and anxious about the event to come. I cycle through my pain calculations. I need to eat but there's no room service. The closest café is two blocks away and I'm not sure I can afford to expend the energy required to walk there. I have only an hour before I have to leave for San Siro. I'm worried about how much walking, standing, enduring this experience will require and as I think about it, I begin to spiral. All

I see is pain lurking in the ever-approaching evening. I turn on the TV and find a channel that's just beginning a James Bond marathon. The movies are dubbed in Italian and commercials disrupt the action every ten minutes. The hotel floor is cool and comfortable. Bond is on the screen, chasing bad guys, running around Europe for me. I can't think of anything I want more than to stay on my back and let hour after indifferent hour flicker by.

I breathe and find my place in the neutral room. From there, I can release my beliefs about the night to come and focus solely on my next task, which is stretching out my spine. I go through my rotations, I count the seconds. I am relieved to be alone. I'd feel self-conscious doing this in front of anyone, not just because I look ridiculous but also because seeing me in pain often stresses other people out. People want to help but don't know how and so they fret and this adds weight to everything. It is easier to hide how I feel, and now with Wolfgang, almost six years old, this hiding takes on new importance. I do not want him to worry about me or take on my anxieties and so I keep quiet to protect him, but no matter what I did he absorbed and reflected back everything anyway. He watched me so closely. He followed me wherever I went, searching my face patiently, clinging to my side, never wanting us to be apart. He'd named himself my *Little Shadow*. I look at the clock and imagine him at home. *Time for bed, my Little Shadow.*

I take the subway to the Lotto stop, then transfer to a trolley that takes me to San Siro, the largest stadium in Italy. The grounds surrounding it have been transformed into a small village of commerce. Hawkers call out from their kiosks to passersby, advertising their wares—commemorative T-shirts, glossy posters, tchotchkes, keychains, bottled water, fried food.

I watch a man hunting buyers for his braided string bracelets. He stalks through the swell of people rushing through the stadium gates. He approaches a young girl, but she ignores him and as she walks away, he places a bracelet on her sleeve and the strings catch on her clothing. She doesn't notice until he begins to shout. He needs people's attention. He needs the eyes of others to make the grift work. Over and over, he accuses the girl of stealing. She sees his bracelet stuck on her and shrieks. She shakes her arm, but the thin threads cling to her sleeve. The man argues with her, demands payment. The girl turns red with embarrassment. People are watching. She looks to them and not the man and that's how he knows he's got her. Her discomfort is his best tool. He continues to berate the girl, playing now solely to the crowd to increase their attention and build her humiliation past the point where it is bearable. The crowd grows, but no one says anything. They just stare and so do I. Finally, the girl lets out a shaky, angry breath and hands over a euro. The man takes her money and disappears. The girl, no older than fourteen, holds her reluctant purchase out in front of her, walks it to the trash, and flicks it in.

The earth steams. White, hot air, thick as glue, plumes from the pavement, filling my lungs until they ache. I peel my dress, wet with sweat, from my stomach and the backs of my legs. I walk toward San Siro, a massive castle of steel and concrete, with the anticipatory dread of personal punishment. I am dismayed to see long lines surrounding the stadium. I'd imagined being able to get to my assigned seat right away to rest, but no one is being let inside yet. I think to leave but I don't want the shame and hurt I will feel if I look directly and honestly at how inaccessible this situation is for me. I do not want to be left out. No one else seems concerned. I listen to laughter and happy chatter. For a moment, I see San Siro through the eyes of the people around me, as a structure intended for pleasure, leisure, joy. We are gathered here to have a good time.

San Siro can fit in eighty thousand people, one of whom knocks me over as they race past me to secure a spot in line. I fall onto the concrete, scrape my hands and knees. Two women appear, mutter at me in Italian, grip my arms at the elbow, locking them to my sides. My mobility limitations require me to use my arms in ways that aren't intuitive to others. I have diminished motor control over my legs and need my arms to help me move and balance when getting up or ascending stairs. The two women tug at me awkwardly, urging me to my feet, demanding I stand, but by taking away the use of my arms, they make it harder for me to do the thing they are so insistent I do.

"I'm fine," I say. *"Tutto bene, tutto bene."*

I try to shake the women loose, which makes them laugh.

"Tutto bene," I say again and this time I don't hide my frustration and the women laugh more. It is hilarious to see me angry. They coo and smile and hold me tighter. It is common for my emotions to be received as non-threatening and childish. My shyness, my small size, and what looks like fragility, all combine in the minds of others as reasons to infantilize me and dismiss my agency.

The women deposit me at the end of a long line. Everyone must wait to enter the stadium. A wave of defeat washes over me, leaving me briefly nauseated. My knee bleeds. A fashionable group of men stand in front of me. They smoke cigarettes and sip what smells like vodka from water bottles. They dance in place, hugging each other. They speak in a language I don't understand, not because it is Italian, but because it is clearly the secretive language of intimate friendship. One notices me standing there and the corners of his lips ease upward into something like a smile. Sweaty, red-faced, bloody in a damp dress—I must have looked like a bad omen. He shudders and lights a cigarette.

I look around and notice other lines are moving. I double-check my ticket to be sure I'm at the correct entry point to the stadium. The smoking Italian sees me looking around with concern and smiles sympathetically.

"English?" I ask. His smile widens.

I show him my ticket and ask if I'm in the right line. He nods and then he points to the place on my ticket that translates to "Midsection GA." I do not have a seat assignment. I have access to a standing-only space in the middle of the stadium's massive open field in general admission. I will not be able to sit. I will be crammed in with taller bodies all around me, unable to see a thing.

A bus idles in the lot behind me. It would be easy enough to walk toward it, to board it and let it take me back to the hotel. Beneath humiliation, I feel relief. Tea, heating pad, pain pills; no lines, no more standing, no strangers. I'd wanted an excuse and I'd gotten it. Then I catch a glimpse of the grifter with the braided string bracelets wandering around the lot, seeking his next target, and a new thought arises.

I turn back, away from the bus, toward the cool, cavernous underbelly of the stadium. I walk toward the security guards at the front of the line. I slow my steps and amplify my limp. I stare at the waiting people I pass in the line. I need their eyes to follow me. The smoking Italian and his friends turn with curiosity and watch me go by. I nod to the guards solemnly as I approach. One opens his mouth to speak but his words evaporate as he watches me hobble toward, then past him. I duck under the flimsy chain at the gate and cross the threshold and am inside the stadium, away from the long line of people who must wait. No one says a word to me. I keep walking.

Two months before I arrived in Milan, I was leading my night class in a review session for their final. It was late May, a few weeks after the night at the bar with Jay and Colin. I hadn't seen either since but turned the memories of our conversation over and over in my head.

I stood at the chalkboard and asked my students what they remembered of Descartes's *Meditations*. We moved on to Kant, then Hume.

We'd reached the point in the class where the students were watching the wall clock above my head instead of what I wrote on the board. A few began to pack up their bags.

Just then Sharon burst through the door cursing. She took a seat in the front row. I paused at the chalkboard. Sharon was older than my other students, starting college over again in order to begin a new career. She wore a full face of makeup and a pressed pantsuit, one of three she put on in rotation. She often arrived late, then sat in the back of the room and fell asleep.

Earlier in the term, after we'd studied Plato's allegory of the cave, I'd assigned an essay asking students to analyze what Plato sees as the burdens of knowledge, and in response Sharon had written about moving from Nigeria to New York to be with her husband. They had three children but had been apart for fifteen years. He'd left Nigeria to work in the States as a driver. Six months after she'd followed him to New York, her husband left her and their children.

At the end of the essay, she riffed on a theory, which went something like everyone was an idiot, always trying to get out of their own caves by themselves. I thought this was interesting and wrote *Wow!* in the margins but docked her points for not connecting this more concretely back to Plato's allegory.

Sharon fidgeted in her seat. I continued with the review session. Sharon's phone rang and out she went. I heard her yelling. I crossed the room and peered down the hallway at her. She moved toward a locker with alarming speed and smacked it. A metallic *clang* echoed down the hall. I stepped back inside my room and closed the door.

Sharon had not returned by the end of class. Her backpack sat, abandoned, and her car keys rested on her desk. I had to lock the classroom but did not want to leave Sharon's belongings in the hallway, so I stayed in the room, graded papers, waited. The minutes ticked by. I stared at Sharon's empty chair. Finally, I got up and checked the hallway, but she was nowhere. Then, in a whirl of movement, she came racing back toward the classroom. She pushed past me in the doorway, gathered her things.

"Are you OK, Sharon?" I asked. She ignored me. "Sharon?" I asked again.

A front zipper on her bag had come undone and its contents spilled out on the floor—pens, mints, makeup. I walked over and crouched to help her, but she swept my hand back from her pile like she thought I might claim it.

"I'm sorry," I said. "I just wanted to help."

"Yeah, I've got it," she said. She tipped forward suddenly, her heels slipping a bit on the linoleum. "I've got it!" she said again.

Swiftly, she tamed the rest of her objects and righted herself and was out the door, but just as she crossed the threshold, she turned her head and said over her shoulder, "Don't forget to count me present."

"Well . . . no," I said.

"What's that?" And she was back in the room, standing in front of me so fast that I shuddered.

"You missed the entire class," I said.

"I did not," she said. Her voice had changed, turned sharp. "I had my stuff right there."

"But *you* weren't present."

"I was present. You saw my bags, I was here. I showed up."

"Look, whether I count you present or absent is not the point. It won't affect your grade. The fact that you missed the final review session is what matters."

"But I made it here. I was present. You have to count me present."

"You don't get counted present just for appearing, you have to actually *be present*."

"What does that even mean?"

"Your mind, your attention."

"Do you think that's ever what you're getting?" She gestured to the chairs where other students had sat, blinking at me from the void, all that evening. "Look," she said. "I'm supposed to show up and I did. I'm standing right in front of you."

"Yes, but you have to *be here*."

"Are you joking?" said Sharon. "Is this a fucking joke right now?" Her voice kept rising. "I'm right here. I'm present. I'm—"

She paused. She did not move. I did not move. Everything was quiet. She leaned down to put her face level with mine. She was very near me now, her eyes peering into mine. She opened her mouth. She screamed.

It was night, past nine, and the academic offices were closed, the corridors empty, the janitors not yet on shift. Sharon's sound echoed down the hall, her breath was close and so was the warmth of her, her pulsing blood, the veins in her face; my blood pulsed, too, my senses sharpened, the room brightened, my hearing traveled out into the hallway, out through the building, into the courtyard, finding nothing and no one.

"I'm going to leave now, Sharon," I said, and I walked out and she followed me, screaming to me, "Are you fucking kidding me? I'm here, Professor, I'm here."

I walked outside. The courtyard was dark and empty. It was raining, but lightly.

"Fuck you," screamed Sharon.

The walk to the subway was four blocks, then three, then two.

"Fuck you!" Sharon walked beside me. "Fuck you, midget."

People on the street stopped and stared. A man came out of the McDonald's on the corner and said, *Whoa, whoa*, and put his hands up, stopping Sharon.

As soon as I got home, I called Kate, who lived in Kansas.

"Wait, what did she say to you?" Kate asked on the phone. I repeated the whole story. "Oh, *Jo*," she said. "Jo" was an old nickname from my childhood. My mother also called me Jo. "What did you do?"

"Nothing, I just got on the subway."

"Were you so angry?"

"No," I said, and the word sounded wrong coming out of my mouth, but it was the truth. I'd only felt numb.

People who met me often asked, "What happened to you?" It was harder to imagine I'd been born with my disability and not been a victim of some accident or illness. That was a story they'd heard before. Someone is just living their lives and then disaster strikes. Many existing narratives about disability follow this plot; the protagonist has a *before* where they are normal and an *after* where they are not. But I'd only ever had this body, my normal body. My sense of self had been formed in an awareness of *other people's* before and after, waiting for the shift from before they were comfortable to after they'd accepted me. Some people switched to acceptance quickly, some never managed it, but I was always looking for it.

If I got angry, this passage took longer. If I kept quiet, withdrew, and waited, people felt less threatened, grew accustomed to me quicker, and then we could begin to communicate. In his essay "The Handicapped," the writer Randolph Bourne says of his own experiences with disability that "the doors of the deformed man are always locked, and the key is on the outside." He continues, "He may have treasures of charm inside, but they will never be revealed unless the person outside cooperates with him in unlocking the door. A friend becomes, to a much greater degree than with the ordinary man, the indispensable means of discovering one's own personality. One only exists, so to speak, with friends."

Kate already knew about the night in the bar with Colin and Jay. She'd been my best friend all my life and we talked every day. Because we had been so young when we met, there'd been no shift between *before* and *after* she accepted me because we couldn't remember a time before we knew each other. We'd always been together. The years had whittled us down to our essences, she was Kate-ness to me and I was just her Jo. When I told her stories of people reacting to my disability it sometimes took her a second to remember I had a body to react to at all.

I asked her to describe this aspect of our friendship and she'd said, "It is wild, sometimes, for me to remember that there was a day you were born and, on that day, we didn't know each other. As far back as my memories go, you are there. I can't find an image of life before you and I can't imagine what life after you could possibly be like. My understanding of my own consciousness is linked to your existence. This is what family is, I suppose."

Sharon showed up the next week on time for the final and sat in the front row. She completed her exam without a word and walked out the door and I thought, *Well, that's over.* I graded her test first. She aced it.

The day after grades were posted, I checked my email and saw Sharon's name blaring at me there in bold. I waited all day to open her message. It read: *I need to see you, Professor! We need to meet. It's important!*

I lowered my head into my arms, which were crossed on the table in front of me. I opened my eyes into the few dark centimeters between me and that hard surface and I thought through all my options.

Andrew coughed. I looked up at him. He held Wolfgang in his arms. He said, "We'll go with you."

I told Sharon she could meet me in the one large office that serviced all the adjuncts. I picked the day and time most likely to yield a populated office, but when I arrived, the room was empty. The semester was over, everyone was gone for the summer. I turned on all the lights. Andrew and Wolfgang stayed outside to play tag on the quad. I could see them from my window. I spread my stuff around to make it seem like I was less alone. I left my schoolbag at one desk, my coffee cup at a second, and I sat at a third. I heard footsteps. I went to the office door and anxiously secured it so it would stay open. The footsteps came closer.

Sharon arrived in the room, an object under her arm. She unfastened the door's latch and closed it behind her.

"Hi, Professor," she said in a hushed voice.

"Hi," I said.

"This is for your son," she said, handing over a box wrapped in paper.

"Did I tell you I had a son?"

"You bring him up all the time in class."

"Do I?"

"Wolfgang," she said.

"Yes, that's right," I said. I placed the box on the floor.

"Professor," she began, and she then was quiet for a long time. She couldn't look at me and I couldn't look at her. I kept my gaze trained on the glass window behind her. I strained for a sound, I imagined a sound, I yearned for a sound that would not reach me through the too-thick panes of glass—I wanted to hear my son, happy, laughing, running freely in the grass outside my window. That want within me had a blunt force, it knocked me back.

Sharon said, "I've been talking to my children and they are very mad at me."

She waited for me, giving me space to speak if that was what I wanted.

She said, "What I called you. What I said."

"OK, Sharon."

"I'm so sorry, Professor." Her hands lay open on her lap.

"OK, Sharon."

"Do you hate me?"

"No, of course not," I said. I did not know how to talk to her or how honest I could be. She wasn't my student anymore. I'd submitted her final grade and she'd passed my class. I wasn't sure where that left us.

Sharon looked at me then as if she could read this exact thought. Perhaps she sensed the strained neutrality in my voice and posture. She shook her head as if that could erase the stiffness between us.

"Professor," she said gently. "I'd like to explain."

"You don't need to—"

"I was overwhelmed by something that happened and I didn't think I'd make it to class at all."

"It doesn't matter now."

"I'm sorry for what I called you."

The memory returned not as a thought, but as a heat that spread behind my eyes, a tightness in my throat.

"The word you used. Do you know why it's offensive?"

"No," said Sharon.

I considered how much instruction I owed her.

"Look it up," I said.

"I'm sorry, so sorry."

"Thank you, Sharon. The semester is over. Let's move on," I said. I looked back through the window. I was hurt and exhausted and I didn't want this conversation to last a minute longer than necessary. I owed her my professionalism, maybe, but not my sympathy. I hardened myself against her words, retreated from her.

"Would you let me explain?"

"Please, I wish you wouldn't."

"It's been such a hard year, Professor," said Sharon.

"I'm sorry," I said. "You just came here from Nigeria, right?"

"Ghana."

"Oh, I thought in your essay you wrote you were from Nigeria."

"All the same, right?"

"No—" I began defensively.

"It's OK," she said.

"I'm sorry. My mistake. Ghana."

"Yes."

"Did you move here with three kids?"

"Yes, yes. I came to be with my husband, and he didn't want to be with me."

"That must have been hard," I said.

"It's complicated, but all for the best. He was very unhappy. He used to drink and disappear for days. He needed to leave us to find out who he really was outside the expectations set by his family. We're all much happier now. He's a better father to his children. This was not why I was hurting that day, not why I was cruel to you. I do really feel sorry for you and don't want to be mean to someone like you."

"You feel sorry for me?"

"For what you go through."

"I feel sorry for what you go through."

"Because I have no husband?"

"No, that's not it."

"Why, then?"

"I don't know, actually. It just seems difficult, what you are describing."

"Please!" she said with a tinge of disgust. "Don't feel sorry for me. I'm good."

"Well," I said. "I'm good, too."

"OK, Professor, good, we're both good. I get it," Sharon said.

Outside, the light changed. I watched it mutate from blue to gold to green through the huge bay window behind Sharon. Emerald-colored clouds shimmered and fanned out across the sky like peacock feathers. A summer storm was coming. I thought of Andrew and Wolfgang under an awning somewhere. And then I stopped thinking of anyone at all. I went deep inside my neutral room. I dulled my interiority until I was confident I could pass through the rest of this encounter as merely a surface. This was right, the right thing to do. Withdrawal from the intensity of the present moment could be a form of agency often mistaken for passivity. I was annoyed, my right hip throbbed, my back was stiff, the memory of Sharon's insult was fresh, my own tears threatened me, the past threatened me; the incident with Sharon was electrified by the past, by all the cruel words others had spoken to me before her. This left me rigid and alert. I imposed a distance between us so that I would not transfer my hurt to her.

Sharon watched me patiently, searching my face. I let myself meet her gaze. She said nothing. It took me a minute to realize what was happening. She could see my refusal to join her out in the open for a conversation and she was willing to sit with me, to give me all the time I needed, to decide if I could reverse this refusal and participate. Time passed and a quiet feeling between us grew, but Sharon made no attempt to hurry us through the moment. I received this as an act of generosity.

"Professor," said Sharon, "can I explain what happened?"

"Yes," I said.

"I was very angry because I'd recently gone to see Beyoncé."

"What?" I said.

"Yeah." She began to laugh a little.

"Beyoncé?"

"Right."

"Go on."

"I went to her concert," she said. "Look," and Sharon pulled out her phone and showed me a blurry picture. In it, she was radiant, she stood smiling like the sun. Her hair hung in loose curls. Gone were the polyester pantsuits she wore to class. In this picture, she wore a tight black dress, heels, and a glittering gold belt. Three women in similarly sparkling outfits stood with her. They were glowing, beautiful. They had taken, Sharon explained, two trains and a bus to MetLife Stadium in New Jersey. She told me how far she'd climbed in those heels up the stadium's stairs to find her seat. She'd squinted down to the stage and there was Beyoncé.

"I felt big joy," Sharon said. "I couldn't stop crying!"

She and her girlfriends had clung to each other, shouting the words to every song.

"You have to see her," said Sharon, then her voice grew loud. "You have to *witness* her. It's an experience. It's *the experience*. THE BEYONCÉ EXPERIENCE."

I sat back in my seat.

"Do you understand? No, no, of course not." Sharon waved me away with her hands in the air. "There's no way to understand it until you are up close. I thought about how lucky I was to see her with nothing between us." Sharon leaned forward. Her face had shifted to reveal a bit of embarrassment. "Professor," she said sheepishly. "Do you believe in epiphanies?"

"Believe in them how?"

"Do you believe a divine voice can speak to you?"

"No," I said, which I thought was true, but maybe it wasn't. Maybe I did believe in epiphanies and divine voices. My uncle Jon—married to my aunt Georgeanne—had told me a story once about how he was trying to fix a toilet in his house and made a costly mistake, damaging the plumbing badly. He sat on the floor of his bathroom, which was getting quite wet, and he wept. He was shocked by how hard he was crying. It was a moment where the past and present collapsed and every one of his failures arrived in his mind as if they'd just happened. This wave of unexpected despair pummeled him, and he didn't know what to do and then he heard a voice. My uncle Jon is a very practical, scientific person. He is not prone to religious or mystical feeling. So hearing a voice in his head as he cried on the floor next to his broken toilet struck him as significant.

The voice said, calmly but firmly: *So. What are you going to do now?*

And he sat up and he thought to himself, *I'm going to fix this toilet*.

And then he got to work.

Sharon says, "When I was there at the concert, watching her, I heard a voice inside me say, *Pay attention*, and when I did, I saw a woman who knew exactly where she belonged. That was her space. That was certain. I could not imagine feeling so sure about where I belonged, about my own space, but then I could imagine it, because there she was showing me what it looked like."

Sharon sniffled. I handed her a tissue and she blew her nose. I held a tight smile. Something was preventing me from hearing what she was really trying to tell me.

"I didn't want to come to your class that night at all. I wasn't angry at you but you brought it out. It was stupid to be arguing with you about what it took to be counted present. I needed to be asking myself: *Where do I belong? What should I be doing? What would it take to feel sure in the moment?*"

I nodded, not knowing what to say.

"The Monday after the Beyoncé concert," Sharon continued, "I signed up for singing lessons. And I thought, *Maybe I'll quit school, maybe this will be the last class I sit through.* But then my phone rang. My daughter totaled our family van. She broke her arm. She's fine, very lucky. Her hospital bill will be very expensive. Anyway, no more singing lessons. In that moment, I felt afraid I'd missed my chance to make new choices, which is not true but felt true. I was hurt. I was coming from that place of hurt when I was yelling at you."

On our train ride home, I handed over Sharon's box to Wolfgang. He opened it. It was a toy truck. We were quiet for a while, all three of us staring at the truck as if it were an artifact from an unknown civilization. Then, too loudly, I launched into all the reasons why I definitely would never want to go to a Beyoncé concert. Andrew listened, watching me closely. I dismissed it as unaffordable and Andrew shook his head, *no*, and gently reminded me that, due to my recent promotion at school, I could now, for the first time in my adult life, spend the money required to purchase a single concert ticket without going into a panic.

"Well, anyway, I don't think I'll be having any epiphanies at a pop concert," I said, scoffing.

"Maybe, maybe not," said Andrew.

"I watch her music videos. I've got the gist of *the experience*."

"You can be such a snob," Andrew said. "I didn't realize."

"Didn't realize what, exactly?"

"How badly you want to go."

"Are you even listening to me?" I shouted.

"Clearly," he said, "I'm listening very well."

But I didn't want to go to a Beyoncé concert, or perhaps I was mistaking fear for preference. I'd heard but not truly listened to Sharon as she explained her experience. I'd felt separated from our conversation as if I were observing it from the hallway outside my office. I was lost in imagining her commute to the stadium. Two trains and a bus, she'd said. Then the walking, the standing, the stairs, sitting in a hard plastic seat for hours unable to stretch. All of this would take too large a toll on my body. *You have to go*, she'd said. *You have to see her. It's THE EXPERIENCE.* But I couldn't go. The site of the epiphany, inaccessible to me.

My familiar defense mechanism was taking over, which was to feel superior while abstracting to theory. I could convince myself that I was above, in both taste and intelligence, the experience of a pop concert. Anything of such mass appeal must be, by definition, lacking—merely facile pleasure, or what British philosopher Bernard Bosanquet called *easy beauty*.

Easy beauty was apparent and unchallenging: "A simple tune; a simple spatial rhythm . . . a rose; a youthful face, or the human form in its prime, all these afford a plain straightforward pleasure. . . ."

Conversely, *difficult beauty*, wrote Bosanquet, required more time, patience, and a higher amount of concentration. Our ability to appreciate difficult beauty depended on our education, insight, endurance, and our capacity for attention. In difficult beauty, one often encounters *intricacy*, *tension*, and *width*. The intricacy of a difficult aesthetic object can provoke resentment and disgust in us if we are unable to resolve and classify the complex elements of the object. Difficult beauty also required us to stay in a state of "high tension of feeling," and it is our own weakness—the "weakness of the spectators," says Bosanquet, taking the phrase from Aristotle—that causes us to shrink from the challenge of difficult beauty. "The capacity to endure and enjoy feeling at high tension is somewhat rare."

A resolving melody, unambiguous lyrics the masses could sing along to, a bit of glitter, lights, spectacle—that was my premonition of The Beyoncé Experience. Blunt, triumphant, easy beauty. An enjoyable experience, sure, but not where I'd find my epiphanies! Despite this, I kept an eye on her tour dates and I walked around our apartment occasionally stating aloud my disinterest in attending any of them.

In June, on the morning of my birthday, I woke to find Andrew and Wolfgang gone. I walked around the apartment alone. Two opposite emotions competed in me. Then, just as I was putting a name to the two feelings, the front door opened. Andrew and Wolfgang handed over coffee and pastries.

Andrew said, "Happy birthday. Do you want a present?"

"A present for me?"

Andrew showed me something on his cell phone screen.

"What's that?" I asked.

"It's a ticket," he said.

"To what?"

"To a Beyoncé concert."

I was confused. Her tour had already come through and left New York. He showed me the receipt on his phone. The concert was in Milan. He'd cashed in our credit card miles and bought me a flight. Andrew thought I needed to leave to figure something out alone, without the eyes of my family watching.

Bosanquet's third term, "width," is more abstract. He writes that difficult beauty has the ability to disorder and confuse us by disrupting our habitual ways of thinking and doing and being. Our habits build up our little house of self-importance and difficult beauty floods this house, forcing the spectator to "endure a sort of dissolution of the conventional world." To find pleasure in this dissolution requires uncommon strength because it is not always enjoyable to be shown how small and silly your house of self-regard is in the face of bigger and more important things, which is almost everything. Difficult beauty does just this kind of pointed showing. To face this and to recognize it as beauty—even as your habitual view of things, your known world, disintegrates—requires that you learn, Bosanquet writes, to "feel a liberation in it all; it is partly like a holiday in the mountains or a voyage at sea; the customary scale of everything is changed, and you yourself perhaps are revealed to yourself as a trifling insect or a moral prig."

People skitter around me in the bowels of San Siro. Platforms laden with stage equipment roll past. Concession stands light up. Workers wipe counters. I walk as steadily as I can inside the stadium. A group of very official-looking people are gathered ahead of me. They wear walkie-talkies and badges on lanyards. Their bodies block an arch that I see leads out to the open center of the stadium. Crates of bottled water are stacked near the arch. I walk up, remove a bottle, and drink. No one stops me.

"*Scusi,*" I say. The people look down at me and then they move apart to let me pass through the arch. A security guard inside the stadium waves and asks me something in Italian.

"English?" I ask.

He wiggles his hand to signify *so-so.*

He pantomimes: *This is a restricted area and I need a wristband to enter.*

I nod enthusiastically.

He cups his wrist inquisitively.

"*Tutto bene,*" I say. I nod as if responding to a question only I could hear.

He tilts his head. I nod again. Then we wait. We're at an impasse. He cups his wrist again, points to mine, asks me, "Where?"

I lift my purse slowly, unzip it slowly, open its mouth, and I peer into as if down a deep well. My performance officially begins. Glacially, I sift through scraps of paper, train tickets, notes, receipts. I flip through every page of a notebook. I excavate coins, lint, a toy car left behind by Wolfgang. I look up at the security guard and smile for a beat too long. I need to get the other people who are standing nearby to notice and add their eyes to the situation, otherwise the grift won't go. I smile and smile and hold eye contact until the guard can't stand it and begins looking at the others behind me anxiously. This gets their attention and they edge in closer. No one knows what to do with me. No one knows what I am. I need their discomfort to last a moment beyond what is bearable. People regarded me with a tension that could tip into disgust. My body was an object that could reveal "the weakness of the spectators." And if I knew this, accepted this, it was a weakness I could exploit.

"Wristband," says someone behind me.

"Wristband," I say.

"Wristband," says the security guard.

"Wristband," I say. "I don't know. Don't know, don't know, don't know."

Being disabled in public reinforces a lesson over and over: people want things to look right to them. To get what I want, I must only play the role of the confused little cripple. It will not look right to hassle me. The security guard searches the gathering crowd. His eyes plead for guidance. It is a mistake in my favor. I've got him now.

"It's OK." The security guard leans down to meet my eyes. He's unhappy and this is my best tool. I hear someone behind me inhale.

Everyone is so tense! I chew the inside of my cheek to keep from laughing. "You lose it. It's OK," he says and he fastens a VIP bracelet to my wrist and sends me on my way.

When I first learned, as a child, what my disability meant to other people, I thought: *I ought to start stealing from the Girl Scouts.*

It was very important to our troop leader that we sell more Girl Scout cookies than our rival troops. She'd send us home from meetings with xeroxed copies of event calendars: home football games at the high school, concerts, recitals, church on Sundays for service and on Tuesdays for AA meetings. She wrote girls' names below each of the events, assignments for the stakeouts. I noticed my name wasn't on the calendar at all. I pointed this out to our leader and she hugged me and told me it would be OK for me to just try to sell to my neighbors or family members.

We turned in our order sheets and the next week picked up the cookies to distribute to our buyers. When I got mine, I saw immediately there'd been a mistake. I'd been given more boxes to deliver than I'd sold. I told our leader this and she smiled a thin smile and said that I should have extras. *Just in case*. I asked the other girls if they had extra boxes of cookies *just in case* and they didn't.

After delivering all my orders, I saved the remaining boxes, hiding them in my closet. I sat inside, with the closet doors closed, eating my surplus cookies, the hems of my dresses brushing across my forehead.

Our troop leader asked me how the deliveries had gone and if I needed help. I said no, but she didn't look so sure. She gave me more boxes, *just in case*, and I stacked them neatly in my closet next to the others. When it was time again to sell cookies the following year, she showed me where she kept the boxes and said to take them when I needed them. She reached out for me, pulling me forward, pressing me uncomfortably against her chest.

"God has given you so many challenges, but He's also put me in your path to help." She released me and her face looked wet. "My daughter and I worry about you all the time, sweetheart, but really, you are an inspiration to us all."

She hugged me like I was the depository for the town's bad luck and because I'd taken it, the town was free, her daughter was free. Some people believed that all the bad luck in the world was searching for a home and because it had chosen me, it would leave them alone. These people gave me their version of gratitude. I helped them see how free their own lives were. I was the bit of dark rain that, once gone, leaves behind the bright, clean morning.

The term "disability" did not help me understand myself but was instead a tool for deciphering the strange and disorienting moments when strangers would look at me and decide who I was and what I could do. People saw contrast between their bodies and mine. They saw absence, lack. But I, having only ever been in my body, did not feel lacking. Going up the stairs feels like going up the stairs. Walking feels like walking. It looks strange, I guess, to those who watch me. It looks lesser. But I had no reason to feel lesser. That would require lessons, for which I had many willing teachers.

People make spaces I cannot enter, teaching me how forgotten I am, how excluded I am from "real life." I was stared at but not observed. I was both in and above the world, seeing from a distance, my consciousness forming in remove.

I felt shame when excluded, like I'd been singled out for this unique punishment, and I didn't know what I'd done to deserve it. But my shame's twin was my self-righteous repulsion for able-bodied people who did not see me as real, who did not try, who were more comfortable holding me at a distance from real life. In the *Republic*, Plato separates people into classes, the highest of which are the philosophers, whose obsession with the futile work of narrowing the separation between experience and truth was what made them noble. Through Plato's lens, I could choose to resee my separation from others as a badge of honor. I could twist this theory into the shape of a shield. Not being of the world was precisely what made me better, wiser, a philosopher, my soul gold and the others' iron. These theories contained in them a superiority, and once I embraced it, it kept me aloft, saved me from further descent. Judgment became a powerful antidote to despair. I thought: *If I must exist at a distance, let it be from above.*

I started to sell the extra boxes of Girl Scout cookies. I pocketed the profit. Soon I saw that I would earn more by breaking up the boxes, putting a selection of three different types of cookies together in a sandwich bag. I called them "variety packs" and sold them at lunchtime to my classmates.

By the third year, I was making up fake buyers on my order sheet. The more buyers, the more our troop leader fretted and over-ordered to compensate for my imminent failures. My variety packs filled an important gap in our school's market. Our elementary school had cookie-hungry kids with lunch money. I could barely keep up with the demand. I had no competition. The other girls had either never thought to sell to our peers or, more likely, knew it was strictly against school rules. But me, I could do it in plain sight. Teachers tried not to notice me. I upped the price.

I kept the money jammed in the toes of shoes I rarely wore. This worked until my mom did a spring purge of our closets. She lined up the pairs of shoes stuffed with money on the dining table. She did not speak. I nodded, acknowledging the situation. She gathered the money and the remaining extra boxes of cookies and walked me outside. We climbed into her truck and drove to the troop leader's house. She opened the door and invited my mother inside, but my mother shook her head and handed over my loot. The two women stood in the doorway talking for what felt to me like a long time. I couldn't hear what was being said, but I could see my mother's face. When she got back in the truck, she informed me that I was kicked out of the troop for good.

I asked if she was mad at me. Nearly imperceptibly she shook her head. *No*. Then we were quiet. We drove through the streets of our small Kansas town, past the community pool where, in the summer to come, I'd stand in line at the diving board and listen while boys compared my body to those of the girls who stood ahead of and behind me; we drove past the tiny library where I'd seek and find refuge; we drove past the junior high school where, later, I'd learn to find comfort in loneliness, and onto the gravel road that led to our farmhouse. Finally, she spoke.

"Some people are going to look at your body and see that something's wrong," she said. "That will be their first and only thought about you: something's wrong. They will not attempt a second thought."

Bosanquet had written that difficult beauty was hard to parse in part because most minds resent "great effort of concentration." My mother didn't want me to expect that effort from other people. Bosanquet went on to qualify it as "not exactly an intellectual effort; it is something more, it is an imaginative effort."

"Look," said my mother, who held no faith in other people's imagination, "just play your card, get whatever you need from those people, and move on."

I can only know now, now that I have Wolfgang, how painful this must have been for her. She wanted to arm me, to make me strong, but this came at the cost of connection. I was always wary, always waiting for someone to be unkind. Both my mother and Colin had described disability as a card in my hand. Colin thought it was all a bad hand I rationalized away. But my mother had always framed it as a tool that could be used to my advantage or, at least, for my protection.

With my wristband, I join a group of VIPs in a small, barricaded area directly to the left of the stage. I stand up against the barricade and watch roadies load in equipment. Now nothing blocks my view. The general admission gates open and the rest of the crowd races in, but I am protected from them in the VIP section. The summer sun holds strong. I'm sweating again. I see down the field the smoking Italian dancing with his friends in their beautiful clothes. I miss Kate and wish she were with me. I miss my mother, too. I could be with them and not feel my mind split into two parts.

The show starts late. My back is stiff, then my hips lock up. An opening act trickles on and then off the stage. People join me at the barricade, press in against me. I push back to make a little space for myself. Stretch my spine, bending myself in half. A DJ emerges. No one is too impressed. We are just waiting. Again, I hold tight to the top of the barricade and fold in half. I listen to my spine crack, vertebrae by vertebrae, feeling temporary relief flood my body. My heels are so sore, my Achilles tendon feels shaved. Pain radiates. I transfer my weight from one side of my body to the other. There is no way to sit, nor leave and come back. I notice the lady next to me eyeing me suspiciously.

"Tutto bene?" she asks and I nod yes. She waits for me to grimace, to confirm her concern.

I look out over the barricade at the open stage in front of me and feel grateful and miserable and annoyed at myself. I'd thought, while standing in line outside the stadium, that I'd had no real choices. I could only either take advantage of people's tendency to infantilize me or be excluded completely from the thing I'd come to experience. I had been so willing to choose to aid people's reductive assumptions about me to gain what I wanted. Wolfgang could never see me do that. I could not do that again. I had been living a life that might have been worthy of me, but it was not worthy of him.

I needed a new way.

That night in the bar with Colin, I'd tried to explain that disability had shaped me in positive ways. This was true, but my explanations had been unconvincing because I lacked the language to explain my relationship to my disability. I never talked about it. I didn't write about it, acknowledge it, study it. I hid it in photographs. I'd spent my life waiting for people to reach their place of comfort with my disability so that they'd forget about it and then *I* could be seen. Of course, I'd succeeded only in erasing a part of myself. The same feeling had dominated my conversation with Sharon. I'd missed an opportunity with her, to learn from her and connect with her. She'd been willing to really talk to me, but I'd been unable to truly speak.

A horrible sound comes out of me, an unintentional sob of sorts, a groan and a wheeze. The suspicious woman next to me whips around and there I am, the shortest woman for miles and miles, bright red, briny with sweat, bent over, back cracking.

"Tutto bene?" she shouts.

"No, no, I'm OK," I say. *"Tutto bene."*

She leans away from me, eyes wide, and then calls for security. A guard appears and she says something to him in rapid Italian while gesturing at me.

"No," I say. *"Tutto bene."*

She stays suspicious and watches me, trying to catch my deceit.

"She needs out, out," she says, pointing to me. The security guard squints.

"I'm fine," I say. "Please tell him I'm fine." But she's not so sure.

People near us stare.

The security guard signals to another guard. Together they move toward me. My face flushes and I feel angry tears eager to appear.

"Stop. Everything is fine. Stop, please," I say. I slap at their hands, which begin to grasp me, hook beneath my armpits. I look behind me. The crowd is watching. Some wear a smirk, eyes wide, faces made gleeful by the possibility of an impending incident. The men are ready to haul me out; their fingers slide down my arms and lodge into the soft skin around my elbows; they pull me up but I'm stuck, bent at the waist over the metal barricade. The men tug, but the encroaching VIPs, giddy with opening space, rush forward and pin my legs. The men curse in Italian at my unmoving body.

"Medico, medico," I hear a guard repeat into his walkie-talkie. *"Medico, medico."*

"I'm fine!" I say over and over, but no one is listening to me.

The men finally wrench me free from the crowd and I pop up and over the barricade. Two men run toward me carrying a stretcher. Was the whole stadium watching me? One of the medics approaches.

"English?" he says.

"Yes," I say.

"We can take you to the medical office," he says.

"I can't leave!" I say. "They pulled me out of the crowd even though I asked them not to."

"But your safety," says the medic. He looks over the barricade. Bodies had rushed to fill the spot where mine had been. There is no putting me back. The medic hands me a bottle of water. He turns back to the security guards and the other medic; they stand in a circle talking, and each takes a turn shrugging.

"You can watch from here?" the medic asks and puts his hand at the edge of the stage.

"Where?"

"Here, on the stage?"

"Watch the whole show from the stage?" My voice involuntarily rises to a shrill note.

"Yes, is that better?"

"Is it better to sit on the stage?"

"For you?"

"Yes," I say. "Sure. That's better. For me."

The lights go down. The crowd screams. She arrives and snaps us all directly into the present. She stands still onstage. The sight of her fills us up. She approves. The show can begin now.

Sound pulses through me, sending my attention shooting from my head out through my limbs, I am pulled forward. The noise of the crowd, their wild pleasure, their big joy, presses in all around me. The pain I'd felt moments earlier is gone, I feel it leave through my fingertips, pushed out by sound. It would return later. Tomorrow in the bath, I'll stare at the pattern of blue bruises the men left on me. But I'd barely remember the discomfort; its effect temporary.

Beyoncé's many powers are on display, but I am struck most by her ability to be inside the moment, to be fully present.

I have never seen anything like it before.

It is as if every molecule of her body is tuned toward us, to be with us, right then. She gives herself to us alone and the audience gives itself to her. When she sings, we sing back with one voice. When she moves, we follow. We are a single organism. A sheet of people. When she goes right, we go right. When she moves left, so do we. When we cannot see her, we sigh in defeat. When she comes closer, to the edge of the stage, we lift our arms up to her, stretch our bodies out toward her, to be closer. We wave and she waves back. Everyone is up on their toes, with their arms extended, fingers unfurled, longing to be nearer, and, out of eighty thousand people, I am the nearest to her. For the next two hours, she performs an act of total generosity, showing us what the absoluteness of presence looks like and—more than that—she brings us into the sphere of this *being here*, and we are, briefly, like her, not above but within the moment.

I look out from the stage at an ocean of people, all united in an experience of blunt, triumphant beauty, and I think of all the ways I'd nearly rationalized myself away from this, of how many layers of superiority, theory, pretense I'd used to build up my little house of self-regard that kept me inside, shielded. I feel ashamed of my weakness as a spectator, my own unwillingness to sit out in the open, to face hard facts, complexity, high tension of feeling. What else have I lost to my defensiveness?

Difficult beauty, Bosanquet says, "simply gives you too much, at one moment, of what you are perfectly prepared to enjoy if only you could take it all in." The ability to perceive and appreciate truly complex beauty requires a willingness to process it slowly, bit by bit. We must not demand it make itself apparent all at once.

I think of Wolfgang and all I'd felt when he was born, that day too complex to be apparently happy, too intricate to process all at once. I feel the shape of Wolfgang's body pressing against my chest. I feel the exact weight of him in my arms. I remember the night I stood in his doorway while he and Andrew played on the floor with blocks. I remember the chasm between us, them in their present, me in my past. It is possible to change. I just don't know how. But I am not unable.

I watch the stage and see what presence looks like, what it feels like, what effect it has on others. For that night, no part of me is split, all of me was there, with all these screaming people, all caught up in the wave of her making, which begins to dissolve a corner of my conventional world, and the model for a new future shimmers in the distance. I try to take it in, bit by bit.

Part Two: The Kestrel

7

The Peter Dinklage Party

Days pass, all the same.

Trash in cans steam in August's humid heat. Petals from rotting blossoms fall from branches and the wind sends them skittering like beetles across the avenues. September brings a sudden chill that strips the trees of their leaves. October's easy skies turn orange with smoke from Canadian wildfires. The last of the leaves die, desiccate on the stem. New York City is loud, stays loud. A bar opens below our apartment and other people's drunken conversations filter through our bedroom window, distort and invade my dreams.

I spend my weekdays at school. Teaching goes well. I'm happy and focused in the classroom when I'm with my students, but when I later sit in my office alone and grade papers I lose all my concentration. I am too anxious. Every line I try to read appears out of focus, just slightly. I'm not quite awake, it seems. There's a buzzing in my ears. I'm outside of my body. My awareness functions like a spotlight, selecting a tiny circle of reality. One day, early in November, I open my office window and look out at Harlem below. A pigeon shits on the pavement, the air is thick and hits me like a wet rag. Truck wheels scream. I close the window. I find coiled strands of my hair on my office floor. My hair is falling out? My mind drifts out the window until a knock on my door startles me back into my body. It's the dean of my school. I'm in trouble. I'd lost time and missed the faculty meeting.

At night, in bed, I hold my phone close enough to kiss it and pull up pictures of faraway places. Mountain ranges, waterfalls, railways into cities, streams, beaches; the world reaches for me through the screen. I can't sleep. I'd left Italy hopeful, so hopeful, but I returned to Brooklyn still myself. Knowing I need to change doesn't make it happen. I'd intended to bring home the happiness I'd felt but I can't hold on to it. I stare above me until I disappear into the black swirling ceiling. The sun rises. It is morning again. Wolfgang scrapes his spoon across the bottom of his breakfast bowl, a shovel on the head of a drum. The smell of yogurt, the sour stink of saliva; tattered light, blunted by buildings; buzzsaws buzzing, screams from the street; feet stomp on metal grates and concrete. Elsewhere, a door exhausts its hinges. Andrew coughs, then sneezes: all explosions, they startle and oppose me. I can't read or write. I can't find a place of rest. Day drones.

I resume the performance of normalcy. I cook, I teach my classes, I attempt timely email responses. I hold my son and read him to sleep, I kiss my husband, I see my friends, I eat at restaurants, I go to readings, I go to museums, I move convincingly through my very good life. But in the dark, while my family sleeps, I click on images of temples, jungles, deserts, lakes and rivers, always far away, as far away as possible. I also start to watch a lot of tennis. It is a relentless sport, played on and on, day and night. Its sounds are rhythmic, boring. Matches with long rallies lull me into a dazed mental space. I listen to the ball bouncing off strings—that satisfying *thhhhwop*—over and over.

My friend Bobby calls from California to tell me his wife has left him. I tell him to pack his things and two weeks later he arrives in Brooklyn with a few boxes and sets up a room for himself in what had been my office.

He sits on the couch eating cereal with Wolfgang and Andrew. The three of them look like a family.

A friend of mine is an editor at a culture magazine and he asks if I'd like to write a review of a Hannah Arendt documentary. I tell him I'm not qualified beyond the fact that I'd read Hannah Arendt, which he assures me is the salient qualification. I see the documentary at Film Forum in the Village and, after, write feverishly on napkins at a nearby café. It is the first time my mind has felt sharp and able since Italy. The following day, I meet the director. She's brilliant, serious, and focused. She understands the space she takes up in a room and occupies it unapologetically. She speaks without ambivalence about her ideas, her intentions. I admire her.

My review of her film sneezes into the void of the internet then evaporates, but also it earns me a pat on the back from my editor and his boss on the culture desk. The Sundance Film Festival is coming up in January, they tell me, and they wonder if I'll cover it for the magazine. They need someone they don't have to pay well.

"Don't you have that friend in Utah you could stay with?" my editor asks me while explaining that any incurred expenses will be mine to pay. I do have friends in Salt Lake City, Iris and her husband, Judd. She is five months pregnant with their first child.

My editor does not assure me that I'm qualified to cover Sundance and even after they ask me to go, the magazine searches for someone, anyone, else to send in my place, but everyone they ask wants to be paid a fair wage. My amateur status and lack of experience mean I have no power to negotiate my rate and this proves to be my edge, and so, admitting defeat, the magazine issues me an assignment and a press pass.

My editor tells me I'll need to file two articles a day for ten days, both of which are due every morning by ten a.m. I am to use my press pass to get into as many movies as possible. Four or five a day.

"But when will I write the articles?" I ask.

"Wrong question," he says. "Instead, ask yourself: When will I write the emails?"

It will be endless emailing, he tells me.

"When you think of this trip," he says, "does an image come to mind?"

"Yes," I say.

"Replace it with emails," he says. "Publicists, agents, festival officials, me, the other editors on the culture desk. Circling back, confirming, negotiating, scheduling—then someone cancels and the emails start again. Emailing like breathing, constant."

The day my press credentials are approved, the emails begin to appear in aggressive numbers like an army marching against me, their subject lines all proclaiming a ***MEDIA ALERT**!*

I'm taunted with links to advanced "screeners" of films that will play only if I possess the right password. I never do. I'm warned there will be limited seating at the "P&Is" or the "WPs" or the "pressers." There are pleas for "curtain-raisers" and threats of embargoes. I can't decipher their meanings but the missives keep coming.

That night, I go to Ditmas Park to meet Jay at Sycamore, which is a flower shop by day and a bar by night. Dried bouquets hang from the ceiling. A desiccated petal falls in my drink. All around us, on every windowsill and ledge, plant life in decay. Right away, I moan about the emails. Jay sips his whiskey, listens.

I clear my throat and read, "Hey, guuurl, just a check-in from your favorite patriarchy-fightin' period company that's been making some major crimson waves."

Jay looks confused.

"I don't get it," he says.

"There's an LOL in parentheses after the crimson waves joke," I say.

"Is this a PR email from a tampon company that's . . . going to a film festival?"

"None of the emails make sense. This one makes the least sense."

He leans over my shoulder and we read together, *I know u n' ur readers be wantin the full scoop on running a creative feminist AF company in Trump's America. V exciting!!!*

"Reading that took something from me," says Jay. "It's drained me."

"They are promising me the Full Scoop."

"Are you working on your dissertation?"

"I will someday."

Now that I'd gotten a full-time teaching position on the strength of my first PhD, I'd lost all my motivation to finish my second.

"Is it still arguing something about Cambodian art?"

"Good question."

"Why do this?" Jay asks. "Why bother with this Sundance assignment? It doesn't even seem possible for you to enjoy the festival or any of the films with all this work the magazine wants you to do."

"Yeah," I say. Graciously, he lets me think in a silence that extends past what is polite. He waits, sipping his drink, without a single outward sign of discomfort. I tell him about the Hannah Arendt documentary and how I'd felt writing at the café after. I think of my father's letter and his question that closed it. *How can I live an authentic life in the present?* When I was writing about the film and, later, when I spoke with the director, I'd felt authentically present. So why go to Sundance? I was chasing that feeling.

"Don't take this the wrong way—I think you should write what you want—but I'm curious as to why you don't study or write about disability."

"We've talked about this before," I say.

"No, I asked you about this before," Jay says. "But you didn't answer. Not that you owe me an answer."

"It's OK," I say. "It's a fair question."

"Do you resist identifying as something? Like, with that PR email. It's terrible. But does some of your repulsion come from the fact the company explicitly names itself *feminist AF*?"

"V exciting!" I try to cheers our beers. Jay rolls his eyes.

"Yes, v exciting," he says.

"Stating publicly that you are feminist is great," I say. "Claiming it for branding purposes alone is gross."

"But where's the line between these two acts for a public entity? When does the act of stating become the act of branding?"

The work of analytic philosophy is to name and define the unnamable and indefinable, to draw distinctions between indistinct concepts. Jay, an excellent analytic philosopher by training and disposition, studies at the intersection of philosophy and cognitive science. His research seeks to better understand the specific boundary in the mind where, when crossed, seeing turns into thinking.

"The line is breached," I say, "when one's intention is to claim identity for the sake of personal profit."

"OK, but your own actions seem motivated by a fear of other people's perceptions of your intentions, not the intentions themselves."

"You know, everybody thinks I used my disability to get accepted into the philosophy program."

"No, not everybody thinks that."

"If you don't study disability, you are a bad disabled person. If you do, people become very suspicious. People start calling you savvy. Can I exploit my identity for my own advantage? I can."

"Sure," says Jay.

"And that's some people's first assumption. There's no line between exploring and exploiting. As soon as you identify as something, people start telling you who you are and what you mean. They put you in a little box and leave you there. The second you say you belong to a community, you speak these imposed limitations into existence."

"You're not forced to internalize these limitations."

"But they surround you. You can feel them. You wear them like clothes."

"OK, but claiming community comes with an obvious trade-off, right?"

"Like what?"

"Like you're less alone?"

My editor calls to check in the night before I leave for Utah. "You'll be fine," he says. "You know what to do."

"I don't."

"Huh," he says, resigned. It is too late for us both.

I show up in Utah like a rock kicked off a ledge; my fall into failure is instant and absolute. I go to the wrong theaters at the wrong time. Seating at the Press & Industry screenings is limited and I don't make the cut. Interviews get set up and canceled. Iris and Judd ask to make plans, but plans are impossible. A day's shape remains mere mirage—emerging, shifting, gone.

Iris and Judd—reveling wisely in their final months before parenthood—invite Andrew and Bobby to come to Salt Lake City, too. They'll all hang out, see if they can get into a few movies, and eat fantastic dinners together while I work. Andrew takes Wolfgang to Kansas to stay with my mother and arrives a few days after I do. When he enters our friends' apartment, he—faced with a week of freedom—lays down on their floor and falls asleep. Even unconscious, his presence shifts my luck in inches. I am granted an interview with the director and two stars of what will end up being the best-regarded film from the year's festival. Our conversation is unremarkable, a regurgitation of pat answers to my dull questions, but at the end of the interview, one of the stars looks at me for a long time. He's handsome like hotel furniture, typical and utilitarian. He smiles, reducing his eyes to glinting incisions, and he asks me if I'm going to the Peter Dinklage party.

The Star's publicist perches on the arm of the couch next to him. She winds her long, black hair around her fist. In her other hand she holds an unlit cigarette. She watches the cigarette, rolls it gently between her thumb and her middle finger. The Star tells me that Peter Dinklage will accept an award from a movie database website for being the most looked-up actor on the internet that year. The Star didn't believe this could be true. The same movie database website told me that the Star was supposed to be six foot five, which I now see is a four-inch exaggeration.

"I'm going to ask you a question," he says, and makes a face I know well. It's the *This might sound bad, but you know what I mean* face. "Who do you really think people have looked up more? Me or him?"

"You know she's going to say . . ." The publicist trails off.

The Star's blue eyes widen. He's been in many films, many of which he lists while waiting for my answer. "Who would you guess? If you had to choose—me or Dinklage?" He pats his chest. His publicist watches the white wall in front of her.

"Peter Dinklage," I say.

The Star exits. I gather my things. There's a wet *thunk* and I look up to see him peering back at me through the window. He lifts his shirt and presses his nipple to the glass. He makes kissing noises, licks the air, waves goodbye.

The doormen of Utah issue a lot of apologies. They are so sorry about the snow, the hills, the icy sidewalks, the puddles I step in. I can't stand on the street for more than a minute before a large, white doorman—pained smile and post abandoned—appears at my side to offer me an elbow. It hurts them, they tell me, to look at me. It hurts them to watch as I struggle along the road, buffeted by the wind that builds up snowdrifts. The doormen hold doors open for me when their job is to keep them closed.

I tell my friends and Andrew to go to dinner without me and I make my way to the Peter Dinklage party. I arrive outside the venue fifteen minutes early to find that, I am, in fact, an hour and fifteen minutes early. It is my mistake, based on the Star's bad information and because I am not an invited guest, and yet the doorman apologizes. Sorry, so sorry, I can't be let in quite yet, but I'll be first through the door when the time is right, so sorry, so sorry.

"You're kind," I say to the party's guard. I ask his name. His name is Tim. "Tim," I say, standing close to show him his great height over me. "I'm always mixing up times, Tim." I shiver miserably for good measure. "It's so hard to keep track of everything. Tim, don't you agree?" He agrees. He reassures me that it would be hard for *anyone* to keep it all straight.

I make good use of the next hour. I stand within sight of Tim. I let snow collect on my uncovered head. He cringes. I pace a bit and shake my legs to indicate that I am tired, sore, and cold. I manipulate him with little thought to his humanity or mine. Later that night, when I try to fall asleep but fail, I will remember the promise I'd made myself in Milan to stop helping others infantilize me. How quickly I'd forgotten it. How poorly I held on to my progress. It is so automatic, so ingrained in me, to use other people's reductive tendencies for my benefit. My increased awareness of this problem had not simply dissolved it. But I don't even notice what I'm doing while I stand outside in the snow, waiting for the Peter Dinklage party to start. The hour passes, then Tim waves me over and I walk to him too quickly. Tim pictures my doom. A "Be careful" escapes his worried, downturned mouth. When I am close enough, I slide on the icy sidewalk toward him and he catches me, walks me past the person checking names off of a little sheet of paper, and into the closest available chair.

"Thank you, Tim," I say. "Thank you so much."

Earlier that day, my friend Iris asked if I could tell she was pregnant. We were standing outside. With her large coat zipped up, she looked merely heavily padded for the weather and I told her so.

"Good," she said.

Without the bulky coat, her pregnancy was evident, which meant that people had begun to stare at her body and ask her invasive questions.

"I feel like I understand you a little more," Iris said. "The way people look at you and assume because you're disabled you'll be nicer than you are. I'm starting to get that now. People see my stomach and think they can just touch me."

People loved to caress Iris's belly without her permission. People told her how she ought to spend her time and what she'd have the energy or will to do. They made assumptions. Once they'd affixed the label *mother*, they found her mind transparent, her will static, her body desexualized. Six years earlier, when I'd been just as pregnant as Iris is now, I'd walked into my favorite bookstore to buy myself a big, dense new release and a clerk took one look at me and said, "The baby books are to your left."

Around that same time, an acquaintance invited me to join a social media mommy support group and I ranted to anyone who'd listen about how appalled I was at this invitation. I rejected the idea that support from strangers—whose link to me was motherhood alone—could ever be meaningful. But I lurked through their message boards anyway, reading but not engaging.

I was first in my circle of close friends to become a parent and yet I was certain they'd supply all the support I'd require, and in this way I set us all up for failure. This was before Iris and Judd moved to Salt Lake City. We were all living in the same town at this time and they committed themselves to my aid. They cooked and ran errands and Judd drove me places once my belly got so big I could no longer fit behind my own steering wheel. Every Tuesday, while Andrew was working, Judd picked me up from school and drove me home.

Andrew and I lived in an apartment accessible only by a flight of crumbling steps. There were holes in the wood and no railing. I was slow on those stairs. But I could handle them. Many times a day, I went up and down those steps. I balanced carefully and lifted myself slowly. Or, if the pain in my hips was too great, I went up or down the stairs on my butt like a toddler. It was important to me, at a time when I could not tie my own shoes, to own this small act of personal agency.

Judd hated to watch me walk up these stairs alone. He gripped me by my wrist and elbow in an attempt to stabilize me, not understanding that the free swing of my arms was a necessary balancing mechanism, and he walked me straight up all those rotten steps. Once inside, he would insist on getting me settled in, hanging my coat, bringing me pillows and a glass of water. If I didn't have a dinner plan, he'd go pick up food and bring it back to me. He cleaned our dishes and took out our trash. I had to tell him I was fine over and over before he'd leave. He did all of this because he loved me, loved Andrew, loved our unborn child, and because he was a man most comfortable expressing this love in small, tangible acts of care. I both appreciated him dearly and resented him for taking away the little things I yearned to do for myself. I would ask him to leave me and drive away, preferably without turning to look at me at all. He would laugh at this, ignoring my requests completely.

My pregnant body drew a lot of stares. The side-to-side sway of my walk doubled. I was in constant pain and my mobility was significantly decreased. My hips locked up and it was hard to stand or walk. My small torso meant that my child rested more heavily on my lungs, making it difficult for me to breathe. Sometimes people watched me walk with anticipatory horror, certain I'd split apart in a stiff wind. Sometimes people saw me coming and felt sorry for me. Or nervous. And some people showed true shock. Everyone was curious. How had this happened? Is it permitted? What would I birth?

I kept reading the messages on the mommy support group. There was a discussion thread about co-sleeping, which suggested that to reject co-sleeping ensured a child's sociopathy and to choose co-sleeping caused certain death. I learned a lot about crib brands, mattresses, swaddle cloths, all of which killed children. Formula killed children but breastfeed too long and you'd raise a child incapable of adult love. These women stayed vigilant. They read up on subjects. They were their child's sword and shield. The world named them mother, protector, and these women drew strength and purpose from this designation. They were encompassed by a category that amplified them. They were their child's safest place. But I was my child's danger. I saw this in the eyes of others, the disgust there sometimes; I saw in the way people watched me, in the fear that followed me—I was to be named not quite *mother* but something else.

My OB believed my hips might separate in my third trimester. My misaligned ball and socket joints could pop apart. He prophesied my immobility, bed rest, lasting damage to my spine. But mostly he talked about how my child might come out incomplete and permanently hurt. *Did you ever ask yourself,* my OB said, *is this ethical?*

Six years before Wolfgang was born, a statue—part of a series called *The Complete Marbles* by the artist Marc Quinn—was erected on the fourth plinth in London's Trafalgar Square. The name for the series comes from the British Museum's famous Parthenon Marbles—fifth-century Classical Greek sculpture fragments that were plundered (or, some argue, rescued) from the Parthenon in the 1800s. What remains of the Parthenon Marbles is partial; many parts of the figures are missing, destroyed by war and neglect. Their broken state does not decrease the institutional or public perception of their worth or beauty. Marc Quinn was in the British Museum and noted the awe and reverence that surrounded the fragmented marble statuary. "It struck me," Quinn wrote, "that if someone whose body was in the same shape as the sculptures were to come into the room, most of the admirers would have the opposite reaction. It was interesting to me to see what is acceptable in art, but unacceptable in life."

Quinn's *The Complete Marbles* are sculptures of disabled bodies that bear an intentional resemblance to the Parthenon Marbles. But Quinn's sculptures are not of broken forms, but of whole forms, whole people, complete bodies. Quinn's statue in Trafalgar Square was of a woman named Alison Lapper, who had been born with no arms and foreshortened legs. The sculpture of Lapper is twelve feet tall and carved from thirteen tons of Carrara marble. Her form is naked and eight months pregnant.

The editor of the *British Art Journal* named *Alison Lapper Pregnant* "a repellant artifact." Brendan O'Neill in the *Guardian* wrote that he'd "grown to loathe the statue" and that its continued placement of prominence was indicative of a society that preferred "victims to heroes." Janice Turner wrote in the London *Times* that the pregnant disabled form—Alison Lapper's form, mine—was inherently subversive, mischievous, and the "most forbidden of female forms." Turner added, "Your mind boggles how, exactly, she got pregnant."

I stared at the internet images of the Lapper statue. She was the only disabled pregnant woman I'd ever seen represented in any art form. I longed to talk to her, to learn from her. Through my computer screen, I experienced the kinship of recognition, the kind I'd looked for but had not found with my friends or with other new mothers. But this kinship with Alison Lapper was too strained, too incomplete. It threw my differences into relief. It illuminated my distance from everyone else around me, and I preferred not to look across it.

Alison Lapper's naked form in Trafalgar Square was, apparently, a sign of a society either expanding or contracting, growing or in decay. It shocked, elated, repulsed, provoked people. She struck critics as curious. But above all, she—the subject, not the creator of the statue—struck critics as "shrewd." Personal attacks landed on her and not the artist. One accused her of "commodifying her armlessness." O'Neill wrote that the politics of identity was about giving up agency and the "acceptance of fate." The mere presentation of her body was evidence of bad-faith political maneuvering, of a person willing to cash in on pity. "Lapper assumed her place on the fourth plinth largely through an accident of birth," he wrote, while the other statues were worthy due to their "self-made destiny." O'Neill believed the other statues "commemorate individuals who transformed themselves in the name of achieving some higher purpose" whereas the Lapper statue merely "celebrates one woman's distorted physicality."

My OB put me through a thought exercise: *Imagine your child is running out into the street and you can't run out and protect him? Or what if he can't run at all? What if he is born like you?* He never explained exactly how my non-hereditary disability might in turn disable my child. It was just taken for granted that my body would produce something incomplete. Even now I don't know why my childhood doctors thought I couldn't get pregnant. They never offered a medical reason. It had just always been explained to me, my entire life, that I wouldn't. *Your mind boggles how, exactly, she . . .* And I simply believed my doctors. These external fears, these false limitations, I learned them so young, they became indistinguishable from my own internal voice.

I never felt good or a part of anything good, not for one second, while I was pregnant. In my mind, always, my OB's voice asking me, *Is this ethical? What if he is born like you?* I was angry and alone, even though I was surrounded by other people. I felt so much of my agency was taken from me. But not on the stairs. I would not relinquish my right to walk unaided up my own stairs.

One afternoon, I started to prepare for my escape from Judd long before his car turned down my block. I said, "I'm feeling strong today. I'm doing fine on the stairs. I can go inside by myself." Judd smiled and nodded and gently said, "I'll just walk you up the stairs." I opened my door before the car stopped completely. "I'll get that, I've got the door," Judd said, sprinting to me. I launched myself toward my apartment, swatted his hands away from my schoolbag as he insisted on carrying it. But he never heard me. He was sure he knew better than me. Maybe he thought I was being prideful, and therefore his dismissal of my request was justified.

I became more and more determined to make him understand. A week later, he drove me home and when the car stopped, I told him, "Don't move."

He smiled and came around to my side of the car.

"Don't touch me," I said. "Don't touch me, don't touch me."

He was startled, then amused. He began to bargain with me. Just up the stairs, he said, just give me your heavy bag to carry, just inside the door, just inside the stairwell at least, it's dangerous for you to go up alone, at least let me follow from a distance, just to make sure I'm here if you fall, what if you fell and the baby got hurt—

I shouted for him to leave until he stopped talking.

This man, who sacrificed his time to help me, had tapped in me a well of rage. Time dilated. I felt uncommonly strong. I knew if I kept standing there, if I tried to explain myself, I would hurt him.

He called out to me as I walked away.

"Drive," I said. "Go."

It took twice as long to ascend the stairs. I shut my apartment door. Judd texted. I turned off my phone. I could be alone now.

I think of this day and this exiling of Judd the whole time I'm in Utah. I watch Judd worry over Iris's pregnant belly, clocking her every move with an anxious eye, and I'm thinking of it while I'm waiting for Peter Dinklage to arrive at his party, which I come to realize is actually a private birthday celebration for the CEO of the movie database website. The award presentation is pretext. The attendees are all friends. They group in clumps and line the walls.

I stand in the middle of the room looking for a bare patch of wall I can blend into. There is no place for me to hide. I am alone, turning slowly one way and then the other, in the center of the room. I feel the eyes on me. I watch them notice me, one by one. I turn and it is like I have set off a wave. Their gazes cascade. I face their stares and the wave recedes, returns when I look away, recedes, returns. Finally, salvation: an open bar at the back of the room and I make my move, but am intercepted by a tall, thin woman.

"You must be burning up," she says. Her voice is unnaturally high and she crouches down to be closer to my eye level. "Don't you want me to hang your coat?"

"Nope," I say.

"Don't you want me to hang it up?" she says again, but slower.

"No, thank you." I shake loose from her grip.

A hero enters with hors d'oeuvres. I slam back lobster salad on a cracker. I ask the cater waiter questions until he starts blinking rapidly, torn in his duty to both circle the room and remain polite to me. I should go home. The arguments for this pile up. I wait in line at the bar and when I'm at the front, the bartender asks for the order of the person behind me who looks down awkwardly at me, not saying anything, not wanting to embarrass me or the bartender by drawing attention to the fact that he has not seen me. I wave my hand up and the look of annoyance on the bartender's face is replaced with one of embarrassment. Then he coos, "Oops, I didn't see you there, lil' lady."

I drink my drink. I miss my friends, my husband. I can leave. Any second now, any second I choose. I notice the tall, thin woman who'd tried to take my coat is standing next to me. She smiles a drunk smile and leans way down to meet my eyes and says, "Are you here with Peter Dinklage?"

I don't understand the question at first, because he hasn't arrived yet. Then I notice another group nearby is looking eagerly at me with the same question in their eyes. "Are you related to him?" says the tall, thin, drunk woman.

A microphone screeches. We are asked to form a semicircle around a podium at the front of the room. I stand at one end. Peter Dinklage has arrived and is standing at the other end. We are directly across from each other. He's dressed in all black. I look at him and he looks at me. We hold each other's gaze briefly.

A man takes the mic and starts to talk. Some people listen but other people watch me and Peter Dinklage. Their eyes travel back and forth between us. The man at the mic reads a prepared speech. Peter Dinklage, this man tells us, is a consummate actor and a wonderful friend.

"When we work with Peter," he says, "we know we are standing on the shoulders of a giant." The tall, thin, drunk woman next to me scoffs and whispers, loud enough to make sure I've heard, "Well *that* is a poor choice of words."

Peter Dinklage steps forward and accepts his award with a shrug and says, "Thank you everyone in this room and everyone beyond this room for, ah, looking me up a lot on the internet." The crowd laughs. He continues for a minute more, expressing thanks and wishing the internet database CEO a sincerely happy birthday, and then he gamely holds his trophy aloft and allows photographers their shot at him.

Admirers loom. A wall of bodies closes around him. He faces one camera lens after another and then moves around the room, shakes hands politely, nods when prompted, laughs when expected. He is the evening's curious entertainment. It seems he's picked a set number of minutes for which he'll endure the scene, but once that timer expires, he'll disappear. The internal countdown appears so clearly on his face that if I move a little closer, I might hear the ticking of a stopwatch.

He walks to the bar to order a drink. The bartender looks right over his head until someone else redirects his gaze and the whole awkward scene that had just played out for me plays out again, only this time I am an observer to it. I feel as though I'm seeing myself, except it is not me but international celebrity Peter Dinklage. I hear people talk to him. "*Soooooo*," a stranger begins, "this snow must be miserable for you to walk through." I stand offstage and watch the actor play my role. Peter Dinklage was born with achondroplasia. This shortens his limbs. I do not have achondroplasia, but I do have a similarly shortened lower body. We are, of course, the two shortest adults in the room. We are likely the two shortest adults at all of Sundance.

As the bartender who has overlooked him apologizes profusely, Peter Dinklage turns and searches the room. Finding me, he smiles a gently exasperated smile and a chill runs through me at his acknowledgment, his knowing that I know what he's experiencing. *Kinship*, Plotinus called it. That tingling feeling of recognition when the soul saw itself in another. *Beauty*, Plotinus called it. *Kinship*. Between me and Peter Dinklage? But, no, I need to cool it. His fame makes him falsely familiar to me and I must work to remind myself that I do not know him, have never spoken to him, and am not able to intuit his thoughts and feelings. We are not the same. Our bodies are not the same. Our minds remain our own. We have different medical histories, ages, genders, jobs, tax brackets. He is a very famous person and I don't know how fame changes the experience of being seen by strangers. He belongs to himself alone and is not like me and we are not the same and this is all importantly true.

But maybe this one other thing is also true.

Maybe, in that moment, if I had walked over to him and said, "When I was pregnant, people were afraid," he would not have needed me to explain any further. Maybe if I had said, "This guy wouldn't let me walk up the stairs alone," he would have nodded. Maybe I would not have to explain this hurt to him.

He knows I'm watching. He walks over to me and takes my hand. "My name is Peter," he says.

"My name is Chloé," I say.

"That's a beautiful name," he says.

I tell him I'm looking forward to attending the premiere of his film, which will happen later that week. He thanks me. Our eyes are level. Looking at him, I feel as though I'm glimpsing something familiar and long withheld from me. In his voice, I hear a native tongue; between our palms, pressed together, I feel two lifelines overlaid. My body is not out of place next to his, his body is not out of place next to mine. I feel, for just a moment, normal. If I hugged him, our bodies would fit together like a double set. No work needed to close the distance. No strain. I step forward but stop. My face flushes red. He is not for me to hug. He is not my mirror, my twin, no matter how much I needed him to be.

There was one other kid in my elementary school with a disability. She was autistic. She and her twin sister, a cello prodigy and suspected genius, were both in my grade. The girl and I were always placed together on seating charts and paired together on projects. We tried to be friends, but resentment got in the way. She was athletic, a fast runner, I remember, but was kept on the bleachers with me during gym class whenever I couldn't participate. If she struggled in a subject, it was taken for granted I was struggling in the same way. We were put in a remedial reading class together in second grade even though my reading level was higher than any of my other peers, even higher than her genius sister, whom she was never paired up with in class. I was her true twin in the eyes of our teachers. We were seen as sharing a general unfortunate nature. Where we could have found kinship together, we found only joint dismissal. We were lumped together, disqualified together, and I didn't want to be disqualified, and so I thought that meant I didn't want to be considered disabled. I would not even say the word when I was young; experience taught me that the label meant only the reduction of my individuality. I learned to resist my identity in a grasp at agency and failed to contemplate how I may have, instead, learned to seek my place in a disabled community, one that might have helped me feel less alone.

The tall, thin, drunk woman materializes behind Peter Dinklage. She wiggles over to be between us, getting down on her knees to be at our eye level.

"Sorry to interrupt," she says. "I just have to tell you that my kids love you in *Game of Thrones*."

"OK," he says in a steady tone. She's too close to him. She takes his elbow, slurs her words.

"But"—she leans closer—"they really love you as the elf in *Elf*."

"Excuse me," he says and shakes her loose, turns away, and disappears into the crowd.

The thin woman faces me, mouth open, then looks back to the space just emptied by the fleeing Peter Dinklage.

"Why didn't he want to talk to me?" she says. Alcohol exaggerates her hurt. I shrug and back away until I realize she expects me to tell her the answer.

"Oh," I say. "I don't know."

"My kids really love him in *Elf*. We watch it every Christmas," she says.

"Peter Dinklage doesn't play an elf in *Elf*. Will Ferrell is the elf in *Elf*."

"You can see why I said the wrong thing," she says.

"I can see that you looked at Peter Dinklage and assumed that he must have played an elf in the movie *Elf*."

"Well, I'm being *so* offensive."

I should go home. I'm tired. I want so much to go home and sleep. She puts a hand on my shoulder.

"Explain to me," she says, "what's the problem with an elf playing an elf?"

"An elf?"

"A dwarf playing an elf. Can I call him a dwarf?" Her tone is softer now. There's something she really wants to know and is sure I can help her. "Why wouldn't he take a role as a Christmas elf when he could bring others so much joy by playing one?"

I look away from the woman and see a few other people listening in.

"How could I answer these questions for him?" I say.

"OK," she says, "just answer for yourself. If it would make people happy, wouldn't you accept a role as a nice elf?"

Later that week, I'll be at the premiere of Peter Dinklage's film and a couple will turn to me and ask how it's possible that his character could drive a car and I'll say I'm not with the film and they'll say, *OK, but how do* you *drive a car?*

A stranger in the crowd yells out, "There goes teeny tiny Peter Dinklage," when the actor takes the stage.

And the day after that, a young man in ski gear will push past me on a snowy street and will turn to shout in my face, "Learn how to walk!" And later another man will stop me on the sidewalk to say, "What's wrong with you?" He joins the many others who have stood in my path and asked me this question.

I leave the Peter Dinklage party. I'm alone again. Outside, it is cold and the snow is deep. I take a bus that mercifully drops me only a block from my friends' house. I can see light shining from the living room. It flickers. There's a fire in the fireplace.

A van pulls up next to me. The driver calls out, "Get in, I'll give you a ride home."

"No, thank you," I say.

"I'm with the festival," and he shows me an official-looking transportation badge. "I'm not some weirdo."

"No, thanks," I say. "I'm almost home." I point toward my friends' apartment.

"Just get in," he says. "I'll take you down the block."

"Thanks so much, but no, thank you."

"I can't let you walk home. It's very icy."

"I'll be careful," I say.

"You can't be careful of ice you can't see. The snow is very high. It's harder than you think to walk in snow this high."

"I'm OK, thanks."

I turn away from his window and begin to walk. He follows me slowly in the van.

"I'll just follow you to make sure you get home safely." He speaks to me in an urgent whine. And then he follows me. His wheels crunch slowly through packed snow and so even if I look away, I can hear him there. My face is hot. All my movements are magnified under the eyes of the watching driver. I see myself projected back through his scrutiny. I am strange to him. If I do slip in the snow, this man's outer form would crack apart and out would come a monster of concern, of pity, of misplaced duty, roaring at me, to tell me, *I told you so*.

The man idles in his van, watching me until I'm all the way inside the apartment. Inside is a room full of people who know and love me. My friends, my husband. They're drinking wine in front of a fire. Bobby pours me a glass. I tell them about the man who'd followed me home and Iris says absently, "Oh, that's nice. There is a lot of snow out there."

Judd looks concerned. "I should have picked you up," he says. "I should have driven you home."

"Do you remember how Judd would chauffeur you around when you were pregnant and you wouldn't let him help you up the stairs?" asks Iris. "You're such a brat sometimes." Judd laughs. Andrew nods.

Bobby turns a log over in the fire. They aren't wrong. I am a brat sometimes. "But now Judd's fussing over everything I do," says Iris. "It can get annoying."

"I just wanted to help," says Judd, his face flushed.

Coat and boots come off. I sit on the couch. My friends tell me about their day. They speak to me across a vast space.

At the end of our night at Sycamore, right before I flew to Utah, I said to Jay, "In our philosophy department, everyone expects me to do feminist philosophy or disability studies, but the people who do that work get written off as cashing in on identity studies and not doing 'real philosophy.' Worse, when I write for magazines, I'm told I'm a writer but not a real academic. But to editors, I'm an academic and not a real writer, a nobody, a last resort, and to journalists I'm an interloper and to agents and publicists I'm a fucking idiot—"

"Yes," he said, "but you're doing all of this on purpose. No one is making you cover a film festival. You keep accepting ways to remain the outsider, the novice. You're choosing it. We're way past talking about disability stereotypes now. I think you are afraid someone will identify you as anything at all. What are you trying to evade? Who are you trying to beat to the punch?"

At Sycamore, my thinking was locked in a false dichotomy: I thought I was either limited by a definition or left lonely outside of one, but I'd not left room for this third possibility, which found its way to me in Peter Dinklage's handshake. He'd given me a moment of feeling effortlessly seen and understood. He'd given that to me in real time and in real space. But it couldn't last. I couldn't hold on to the feeling. Eventually the rest of the world crowded in, the world that constantly reminds me I'm out of place, suspicious, and strange; a subversive, mischievous, forbidden body; a forgotten body, overlooked in the shape of chairs, of high counters, sinks, stairs, shelves. But because everything is this way, I can get used to it and can even convince myself not to ask for change. I can go numb. And in this way, I can disappear, even to myself.

I sit in the glow of the fire, in the home of my friends, in the arms of my husband. The distance between myself and them has never been greater. I feel unknown to them, alone. There is nothing that will change this. I can tell them my stories, but I can't transfer my experience. This had been easier to ignore before shaking hands with Peter Dinklage. But what good was his understanding or the feeling of being understood if it made everyday life harder? Numbness let me live among nondisabled people. Kinship held me up for only a moment and then dropped me from a greater height, leaving me suddenly more alert to every ache, reopening every wound; I feel my palm, I pull away from Andrew, I put my arms around myself; I feel only the constant strain of translation. He could love me but not know me. What was I trying to beat to the punch? What was I evading?

8

Static

I am home two weeks before I buy another plane ticket. I tell my dean I have to do dissertation research and ask to teach a few weeks of my course online. I'm on thin ice with her. "You'll miss another faculty meeting," she says. She loves to scold, and the faculty meetings are the stage from which she scolds us best. I endure these meetings with a petulant, outsized agony. I smile at the thought of missing another one. Her eyes widen. Committee work is a significant part of my full-time job, she reminds me. She needs me to review more accreditation paperwork faster. I can't bring myself to acknowledge this. She sets her jaw against me but grants my leave request.

That I might be close to losing this job, the source of my family's stability, does not scare me as it should. I'm too low, as low as I've ever been.

Andrew gently raises an eyebrow when I tell him where I'm going.

"OK," he says. There is distance between us, but this does not appear to worry him. He is not afraid he'll lose me. He drives me to the airport and kisses me goodbye. "Call me when you land," he says.

It takes two flights and twenty-four hours to get to Cambodia. I sit in the airplane's cramped cabin, guarded on all sides by sleeping people. It is dark except for the TV screen embedded in the chair in front of me. I blink at the little square universe there. My eyes grow heavy but won't stay shut. We begin our descent. Across the aisle, someone raises their window shade and I see the world below blurred to blue opaline. Then, without warning, the lights click on. Flight attendants march us up and out of our contained spaces and expel us into the vast and dizzying brightness of day.

Instantly, I'm overwhelmed by the noise, the heat, the pulse; all around me, mammal and plant life alike are overexposed under the blown-out sun. I shield my eyes and try to adjust. A crowd of men shout, "Tuk tuk." The chorus continues, "Tuk tuk, tuk tuk, tuk tuk." It's not me they want, it's anyone. A short path between us ends with the men crouching like fishermen downstream, arms outstretched like nets, ready to catch any one of us. People push up behind me. I have nowhere to go but toward. Swiftly, I'm surrounded. I feel bodies, rigid with urgency. The men, by clamor, compel me to understand: *I must come with them.*

A hand grips the strap of my backpack, lifts it from my shoulder. Someone has it now, has everything I need—my clothes, my camera, my wallet, my passport. The man carrying it walks away. I follow him. He leads me to a covered cart hooked onto the back of a motorcycle. "Tuk tuk," he says, pointing at his rig. He leaves my backpack on the seat, grabs me by the elbow, and tries to drag me up and in. The man's hand grips me tighter. This wakes me from a sort of slumber. I decide not to be forced into his cart. I take my backpack and walk away. He follows me, saying, "Tuk tuk!" but I shake my head. He turns back, away from me, to see if he can snag some other tourist.

The voices of strangers keep calling out to me. "Hotel? Hotel? Tuk tuk to hotel?" I keep saying no and walking forward. I push through the throng, rejecting offers, breaking eye contact, until I'm free and alone, outside of the crowd, where I am forced to face my error: I do need a tuk tuk. I look back at the mass I've just escaped and imagine the pain of reentering it.

Across the airport's lot, a young man stands alone. He regards me sleepily. He, too, is a tuk tuk driver, but he's separated himself from the others. He doesn't fight for anyone's attention. He waits. I look back behind me and then again at him. He leans against his motorcycle. He doesn't move at all until I take a step toward him, then he extends his hand and I walk to him and take it. His name is Chetra. He lets me use his arm and shoulder for balance as I climb into his cart.

The airport shrinks behind us as we charge into the heart of Phnom Penh. The air smells like smoke. We pass many small fires burning. The dust from the road mixes with my sweat and a film forms on my skin. Motorcycles are everywhere—moving with traffic, against and across it—adhering only to the law of the immediate present. There are whole families piled on a single motorcycle, balanced like stacked acrobats on a single slim seat. A boy Wolfgang's age sits on two inches. He holds his legs out wide to not burn himself on the exhaust pipe. A wrong bump would fling this child like a flicked bug into the onslaught. I watch him, my breath sealed in my chest, as his father swerves around other drivers, up and over a small median, and then across another heavy lane of traffic. There's a thud and Chetra's cart lurches forward. We've been hit by a Honda scooter. A young woman is driving it. A baby sits in front of her, its tiny hands resting on her forearms. The young woman's mouth is covered by a medical mask. She stares straight ahead. Horns blare. She wrenches the Honda's handlebars back and forth to stop her wobbling scooter from toppling.

I want to go home. Why have I come so far alone? The air is thick and hot. I sweat and sweat some more. We hit a huge pothole and I go airborne. My body flops hard on the way back down. I cross my arms against myself and make myself smaller. Nothing but this feels familiar.

Wide lanes narrow the closer we get to the city. Traffic slows. It's lunchtime and long lines curl around corner food carts. On a billboard, I see an old publicity still of Leonardo DiCaprio pasted onto a drawing of dancing teeth. It's an ad for the dentist's office below. A picture of Brad Pitt from the nineties hangs above a cell phone store. We arrive at my hotel.

Chetra hands me a placard with pictures on it.

"You choose," he says.

The placard, fat and laminated like a diner menu, presents images of tourist attractions. There's a picture of the Royal Palace, another from the National Museum. The other images are of skulls—a pillar of skulls, skulls in glass cases, skulls on cave floors. Which skulls did I want to visit? I was here to see them all.

"Just point," says Chetra, instructing me on how to order up the next day's experience. I tap my finger on a picture of a dead body chained to a bed. Chetra gives me a thumbs-up and tells me he'll meet me in the morning to take me wherever I want to go.

The concierge greets me in Chinese and then, scanning my face for clues, switches to English. Chinese tourists outnumber other international visitors to Cambodia. All the signs in my hotel are in Chinese and the lunch buffet in the lobby serves only Chinese food. The concierge asks me how many room keys I'll need. I hold up one finger.

"Just you?" he says.

"Yes," I say.

"You have family here?"

"No."

He hands me a key and a list of recommended restaurants in the area. He circles one he is certain I will like the most. "Go here," he says. "This is your place."

I sit on my hotel bed and try not to sleep. I should stay awake at least until sundown to orient my body to this new time zone. Day in Cambodia is night in New York. Andrew asked me to call from the hotel, but now I worry I'll wake him. I worry a text will do the same. I begin an email, but stop, knowing he'll worry when he gets up, looks at his phone first thing, and sees the text he expects from me isn't there. I weigh the weight of every worry and then, before I can decide how to contact him, I'm asleep. I wake in the early darkness of the next day.

From my window, I see nothing. But slowly, fires light. Somewhere, not far away, cart wheels creak and pans clang and scrape; street vendors prepare for the day. The scent of smoke. A motorcycle passes. Its headlamp breaks a brief path in the dark, startles a stray cat, displaces it from where it slept on the sidewalk, sends it into the safety of the alley. The city starts, it warms up with single notes—a horn honks somewhere, an engine mutters, one person whistles below my window—but soon the sounds multiply. Three horns honking make a chord. The accumulation erases separate sounds until all I hear is one grand screeching of machinery, people, life.

Allegedly, I have come to work on my dissertation, the subject of which was ill-defined. The idea was to consider Aristotle's paradox of tragedy through the lens of dark tourism. Aristotle's paradox asks this: If we do not wish to encounter human suffering in real life, why do we take aesthetic pleasure from encountering it in art? Aristotle asks, why did the Greek theater fill with people eager to watch Oedipus claw his eyes out? What is the nature of such pleasure, and did it have a use? One answer comes from Aristotle's *Poetics*, where he defines tragic structure as one that arouses pity and fear with the intention of accomplishing a catharsis of those emotions. Aristotle's vague concept of *catharsis* is often interpreted as a sort of purge, a remedy for the rise of our own negative emotions that, if not experienced and expelled in the safe and neutral space of the theater, may plague us or cause us to act against reason.

Other philosophers have explored this paradox of tragedy by applying its questions to the enjoyment of sad music, grotesque paintings, horror films. I planned to follow this tradition and apply the paradox to people who spent their money and leisure time visiting sites memorializing death and disaster. So-called dark tourism was big business in Cambodia, where a quarter of the population had died in the genocide carried out by the Khmer Rouge in the 1970s. Chetra's menu of tourist attractions mostly offered up proximity to tragedy—prisons, killing fields, torture caves. My plan was to go and observe how visitors interacted with these sites. I anticipated a kind of cross-cultural gawking, and this filled me with a premature ill ease. But was there something meaningful that could be gained in these spaces? Could they lead others to some true insight about human nature that made it ethically justifiable, advisable, good? I thought, in the four weeks I'd be in Cambodia, that I'd try to find an answer to these questions and, in answering them, I hoped to gain a better understanding of Aristotle's paradox of tragedy or, at minimum, I hoped to gain a draft of an academic article pretending to better understand it.

Of course, there was the other side—the self-serving side of me that nurtured the naïve belief, as people often do when they travel far from home, that I'd also stumble upon some better understanding of myself and the distance I felt always, and the resentment I felt less often, toward my normal life. Is it possible that I, through catharsis, will be transformed?

Chetra meets me in front of my hotel and drives me to the gates of Tuol Sleng Genocide Museum. I pay Chetra for the ride and say goodbye, but he shakes his head.

"No," he says. "I'll wait for you." I take my place in the ticket line. He parks at the edge of a group of other tuk tuk drivers who are also waiting outside the prison for their fares. Some sit together playing cards. A food cart approaches and quickly the men bring the food back into their closed circle and eat with their backs turned out. Chetra does not speak to or join in with these men. He leans back to nap, but I see he's not sleeping, he is watching them.

Tuol Sleng had once been a high school. Tall buildings, filled with classrooms, formed a square around a lush and peaceful courtyard. In 1976, the Khmer Rouge converted the high school into Security Prison

21. Its classrooms became torture chambers. In exchange for five US dollars, I get an entry ticket and an audio guide. A recorded voice tells me I am now a living memory. I'm not sure what that means. Then the voice warns that what I'll see inside the walls of the prison might be too terrible for me to handle and that it is permissible to stay outside in the courtyard.

There's a group of Australians ahead of me, speaking softly to each other, waiting to enter prison building A. They wear identical outfits—blue jumpsuits, black sunglasses. They move calmly, as one; and, in eerie unison, they stop speaking and place their headphones over their ears. Presumably, their audio guide is now telling them they, too, are living memories. They pass by boards with pictures of the prisoners and appraise these images with detached solemnity. The guide in my ear tells me that Pol Pot's regime viewed city dwellers and intellectuals as impure. Doctors, nurses, academics, teachers, researchers, engineers were brought first to S-21 to be interrogated and tortured before being transferred outside of the city to the Choeung Ek extermination center, referred to commonly as the Killing Fields.

I trail behind the Australians from cell to cell, but I'm slightly ahead of them on the audio tour, which details the way people have been hurt in these small rooms. I listen and then wait to watch the Australians react. My guide recites a list of human horrors that feels definitive, an absolute archive. The Australians blink, hearing the list for themselves. The guide draws my attention to brown patterns on white walls. *Here you can still see the blood of the dead*, the voice says. The Australians, all synched up, turn to the wall. They observe, nod, take pictures. There are ten of them, ranging in age I'd guess from fifteen to fifty. None cry, none exclaim, and neither do I. We politely listen and witness.

The guide voice tells me that if I'm getting overwhelmed, I can step outside. *Try to find a nice bench in the shade*. And I do want that, so I do step outside, into the quiet courtyard, and I walk beneath the tall trees. The wide leaves of palm trees provide me cover; they blot the harsh light, and darken the manicured paths that lead from prison A to prison B to prison C to prison D. Soon, the Australians exit, too, and settle on the lawn. They bring snacks out of their bags and eat. I walk the perimeter of the quad, stop to pet a stray cat sleeping in a sun patch. Birds sing in the trees. The noise of the city is muffled by the prison walls. The divide is startling. Out there is dust, trash, action, but in here the world is sealed and clean, and roses grow in the gardens. I pause on my walk to admire the simple beauty of the carved wooden arches that extend over the garden. I read the attached placard and discover they are gallows.

I approach the Australians. I tell them I'm a researcher and I'd like to ask them some questions. They respond in a chorus of *Of course*s. They offer me water and a place with them in the shade. I tell them about my project and ask if they could tell me why they're here and what they hoped to gain by coming.

"Just to learn," says the oldest male in the group. "Just to better know this history."

"What do you learn by coming here?"

"The facts of what happened."

"But you could read the facts from anywhere. Why come here in person?"

"Curiosity."

"About yourself?"

"No," he says, grimacing at this strange suggestion. "This doesn't have anything to do with me. I could never do this to people."

On the way back to the hotel, Chetra stops suddenly in the street.

"What are you doing?" I ask.

"For the rain," he says, and he leaps off his motorcycle, comes back to the cart, and unties rolled-up plastic flaps, securing them over my open windows.

But the sky is cloudless, blue. The air is sweet and clear. I sense no sign of rain at all. We wait, but nothing happens. Then I see the sun retreat, dimmed behind massing, ashen clouds that have appeared like magic. The sky breaks open and dumps down heavy rain. Chetra stays outside the cart on his motorcycle and, in seconds, is soaked. I regard him anew, awed by this psychic driving me. But I forget to ask him to sit inside the cart with me. I watch him from cover. He seems fine, and maybe he is even smiling, and so I assume, without questioning my assumption, that he's used to this heavy rain and being stuck out in it must not feel the same for him as it does for me.

When the rain stops, we get back on the road and on our way to the hotel again. But then we discover a new delay. We're stuck in the middle of an unmoving lane. People ahead of us leave their motorbikes and walk up the street, investigating the holdup. Chetra turns back to me and shrugs apologetically. The air carries the scent of ozone and the roads are wet. My life in New York reappears to me, weaves in and out of the air ahead of me. I hear Andrew's voice complaining about drivers in the rain and then he is beside me and we are not in Phnom Penh, but stuck on the Brooklyn Bridge in a rainstorm, watching the water move below and above and all around us, watching people roll down their windows to smoke, watching a man circle the halal cart he's towing, checking that it is securely attached to his truck, and there's Wolfgang in the backseat, laughing, mimicking with his little voice our own annoyance, our admonishments of the cars that honk at gridlock, *Yes, yes, that's sure to fix it*, and I can almost hear his voice cut through the horns in Phnom Penh, Wolfgang's little chirp, sharp with unpracticed sarcasm, aiming to please me, to make me laugh by parroting, shouting out his window, *Yes! Please keep honking. That will help us.* I never minded being stuck in the car with Andrew and I don't mind being stuck now, cozy and alone in the back of Chetra's cart with plastic drapes down around the windows. It's like being tucked away in a little glass case, a dry place from which to watch the world go by.

I see something strange. A pale wave flowing over a distant slope of green. It is people, dressed all in white, moving together from a temple high on a hill down toward and then into the street. They fill the seams between the motorbikes: a great pitcher of milk spilling. They're running toward us. Chetra was, a minute ago, right next to me, but now he's gone. The people in white stop and surround something up ahead. They all look down inside the circle they've made. Just then, Chetra returns and he drives the tuk tuk up and over the sidewalk and we ride past the traffic and I see, at last, the obstruction. It is a man on the ground. Beside him is a smashed-up food cart. A few of the people in white have come so close to this man, dead in the street, that the edges of their white shoes have turned red.

When I return to the hotel, the concierge greets me and reminds me of his dinner recommendation.

"So close, this place, a place for you, so close, you can walk right there!" he says.

His cheerful voice is abrasive and so is the texture of my skin, covered in dust from the road. I return to my room and shower, willing the whole day to dissolve down the drain. It doesn't. I'm anxious. I leave the hotel and go back out into the city, hoping the noise will distract me from myself. Parked in the grass in the courtyard in front of my hotel is Chetra. He's swinging in a hammock he's slung diagonally across the interior of his cart. I'm surprised to see him and surprised by how happy I am that he is there. I call out to him and he looks up from his cell phone.

"Where now?" he says. "I'll drive you."

"I'm walking to dinner," I tell him.

"Where?" he asks again, and then makes a face when I say.

"Not good?" I ask. "Where should I eat?"

"Get in," he says. And just like that I'm back in the cart.

Chetra takes me to a small restaurant on the banks of the Mekong. He knows everyone inside. A waitress walks us through the dining room to a patio in the back. We are the only people out there. The waitress says something to Chetra and points to me. He shakes his head.

"What's she saying?" I ask.

"She thinks you're Australian," he said. "Only Australians come all this way."

"Americans, too," I say.

"Sometimes," says Chetra.

He teases our waitress in Khmer and she blushes. He turns to me and translates. "My cousin," he tells me. "I tell her she's too skinny."

She brings us plates of rice and curry.

"Prahok," says Chetra. He reminds me of the piles of white fish we'd seen on the side of the road and tells me those fish are fermented into a paste that is the base of the curries. I taste it and my mouth and nose fill with the dung musk of long peppers and river mud. It's acrid, earthen, rotten, delicious. The last of the sun melts over the Mekong.

"Were you waiting for me?" I ask Chetra.

"When?"

"You were outside the hotel?"

"The hotel lets me sleep there."

"They give you a room?"

"I sleep in the hammock in the tuk tuk. The hotel yard is quieter than the street. And I'm there to take you where you want to go."

Chetra tells me he has deals like this with many of the major hotels in the area and can sleep all over the city, wherever he chooses. He's come to Phnom Penh from a village in the Prey Veng Province, where his parents live and where they are raising his brother's four children. He sends his money home to them.

"My English is good, so the hotels like me," he says. "And you feel better."

But I hear in his tone and see in his eye the word *safer*. And I did feel safer with him. I'd trusted him immediately on deep instinct. I thought what I'd seen was a kind of similarity between us. We both held ourselves back at the edges of crowds. He was patient while others clawed at me and shouted. But even after one day here, I was coming to understand that what I'd read as aggression from the other tuk tuk drivers was a mode of communication with which I was simply unfamiliar. And so my trust in Chetra had come from confusing the familiar with the good. Maybe it was just that he spoke English and that he knew how to shift into a more Western mode, erasing what part of himself might be less palatable to the strangers who paid him. I saw now in the way he interacted with his cousin and the others he knew in the restaurant that he played a savvy role with me, allowing me a tour through his country at what he recognized was a preferred remove.

"Did you learn English in school?" I ask.

"Some," he says. "More from Facebook." He pulls out his phone and shows me the white faces he follows. He types in my name and finds me and then, with a click, we're friends.

Chetra agrees to let me pay our bill, but only after I've agreed to a trade.

"I'm your Khmer tutor now," he says and, on a napkin, Chetra writes out Khmer phrases he thinks I should know. *No, thank you* and *I don't want it* and *Leave me alone.*

We're invited to stay and have a drink under the restaurant's awning. Chetra's cousin brings us something clear and strong. His friends come outside and pull up chairs around us and talk and laugh with him in Khmer. I turn to the river and watch as moonlight draws dividing lines in the water, cleaving wave from wave. A cat appears and climbs into my lap. Chetra's cousin comes back to refill our glasses.

"Who were those people in white?" I ask. "The people in the street?"

"Which people?" he says and I'm surprised.

"The people all in white? Remember?"

They'd arrived like a dream. They looked to me like a cult, a bad and eerie omen, but I do not describe them this way to Chetra. *Oh*, he says finally, remembering them. He explains that Cambodians wear white when someone dies. So, the people I'd seen in the street must have been coming from a funeral when the accident happened, which meant they'd moved from the observance of one death to see another. One of Chetra's friends elbows him, curious to know what we're talking about. Chetra explains the accident in Khmer and in English. A collision with a food cart, a motorcycle; one man hurt, another dead.

"It is very common," Chetra explains. "Happens every day." Chetra nods his head and his friends do, too, and they start talking again, all at once, but then Chetra's cousin says something that turns Chetra's mood. I scratch the ears of the cat in my lap and wait. Chetra's cousin touches his shoulder and he flinches, just slightly, and he hunches forward over his glass on the table, closing himself off from us. He drinks and his cousin fills his cup again. She turns to me and says something in English, but I don't understand and instead of asking her to repeat what she's said, I smile and nod and—in a decision that will horrify me later—I force out a little laugh.

Everyone is quiet. The air between us is significantly changed. I don't know what I've missed. When his friends finally start talking again, Chetra stops translating for me. My lap is wet. The cat there looks up at me. She turns over and shows me her stomach. She has new kittens nearby. She's nursing, her nipples swollen and dripping. I pet her belly and she purrs. I feed her rice from my plate.

"What will happen to this cat?" I ask.

Everyone looks at me. Chetra's cousin walks back inside the restaurant.

"It is a street cat," says Chetra. "It will die out here eventually."

I think of my father in Asia at my age. I think of the dog he found in his yard. I wonder if I could drown this cat on my lap if I had to and

I know I could not. I am struck with sympathy for my father and I miss him. My face clouds over, I hold the cat to my chest. Chetra, seeing this, stands up. He takes the cat from me and carries her out. No one says anything. When Chetra returns, he's sweating, angry. I'm unsure of the precise nature of my insult, but I've made one all the same.

"Sad about a cat who might not live but not about—?" Chetra begins, but I interrupt him with my apologies. "Why," he says, cutting me off now. "Why even bother?"

The next morning, I meet Chetra at dawn. He's crouched on the seat of his motorcycle, feet tucked up under him, scrolling through something on his phone. I point to the tower of skulls on his tragedy menu and off we go, out of the city and into the country, to the village of Choeung Ek to see the Killing Fields. It's a long ride and we have to stop for gas. Chetra tells me of a fish market nearby and asks me if I want to see it on the way back later.

"I'd like to see it," I say. "But I don't want you to have to wait for me."

"I'll wait," he says.

"You're sure?"

"If I don't, someone else gets your fare."

Streets turn to dirt. The city sinks into fields. We nearly hit a dog with the back wheel of the tuk tuk. There are cats and dogs everywhere. Chetra doesn't swerve to avoid them. He shrugs. There are too many. Killing one is inevitable. I remember my father telling me that dogs and cats roamed the halls of the hospital in Bangkok where I was born. That seemed untrue, but now less so.

The scene outside Choeung Ek is the same as outside Tuol Sleng. White tourists wait in line to get inside while the tuk tuk drivers group together and wait by the gates. Chetra takes a place on the periphery. One of the other drivers points to me and says something in Khmer to

Chetra. Chetra looks at me and back to the driver and says something. I want to ask what is being said, but I've already joined a line that moves me forward. I buy my ticket and enter.

I'm given another set of headphones. A new audio guide leads me. He tells me this is where the Khmer Rouge killed more than a million people. The guide says to follow a path and I do and the path takes me past the glass case filled with femurs, past the pit where women were raped before execution, past the pavilion where people were lined up and shot in the head. I observe other tourists from a distance. I watch them clump around a murder pit, look in. I watch them look for blood in the pavilion. The walking path is smooth and cleared of all debris. The grass on either side is trimmed to an inch.

On the phone with Andrew the night before, I'd told him that I—only a few days into this trip—felt ready to give up my purported dissertation research. I'd discovered only one thing, which was that I'd come to Cambodia with the wrong idea in mind.

"You'll have another idea," Andrew said.

The thing I kept thinking I might discover never materialized. I'd neither witnessed, nor felt myself, anything approaching a cathartic experience. People did not seem to leave these places changed at all. They entered stoic and respectful and, after the requisite taking of pictures and reading of placards, they left stoic and respectful. Or maybe that was just me. Maybe that was just how I felt and I only took note of what confirmed my feelings.

"Well," Andrew began, "as you're talking about this, I'm thinking about what my own process would be like and why I, too, might feel so detached."

"What do you think the answer is?"

He said, "These memorial sites, maybe, induce a mental conflict. We are meant to understand the scope of the suffering while we also know we can't understand the scope of the suffering. I might shut down. I might feel numb, unable to hold it all at once."

The experience, he thought, was cognitively dissonant. And the mind, when faced with dissonance, dismissed. This I knew well.

"Being there," he said, "might feel like moving through a big object made of static. Your mind tries, but can't turn static into meaning and so, it shuts off."

"That helps," I say.

"Helps you but not your research."

"Are you mad I'm gone?"

"I understand you felt like you had to go."

"But it's taking a toll on you."

"I support your decision *and* it is taking a toll on me. Both things."

Andrew did not always act as my defender. He did not think I needed defense from others or from myself. He did not tell me I was right when I wasn't or that things were OK when they weren't. He did not always agree with me or what I did, but he could view things from my perspective without judgment. Such a power was hard to fathom. He brought the full force of his empathy to his every conversation with me. He could do this because he kept his commitments to other people to a minimum. He was pathologically disinterested in status or pleasing people on a large scale. He knew this often worked against him—made him seem hard to get to know, made him appear boring or icy in social situations. Some people dismissed him quickly, but he barely noticed this and cared not at all. He was uncompromising in his priorities. He had chosen to love a small group of people—a few lifelong friends, a few family members, me, Wolfgang—and he gave us the gift of his full attention and energy. He and my mother were the same this way and they loved and respected each other very much. They could be taken at their word, counted on, trusted always, whereas I spread myself too thin, wanted too much, and disappointed people.

I walk for an hour around the paths at Choeung Ek, feeling as though I'm moving through static. I look at placards and listen to my audio guide. I follow the manicured path until it leads me to a fork. One way leads to the last stop on the audio tour. I look ahead and see a large group of tourists already gathered there. This last stop is the big finale. Even from a distance, I can see the rapt expressions of the tourists listening to the last of the audio guide.

The other way leads through a copse of trees on the edge of the field. I take off my headphones and walk toward the trees, eager to be under the protection of their shade. This way I've chosen means walking a long loop and halfway through, I realize I've made a mistake. The twinge in my right hip is sharp and my back is locking up and I'm far away from any places to rest. I retreat into the neutral room in my mind and I begin the process of erecting its white walls. I find my place of rest within them. I breathe in and out. Then I start the count. Numbers flash inside my mind. I narrow my focus to only these numbers. Other feelings slide to the side of me and are left there. I breathe and count. Ahead, there's a chain-link fence I can lean against for a moment. I'm only eight steps away and then eight steps more and then again I count to eight and then again. I keep counting. I forget where I am. I am only inside the room. I reach the fence and grip it for support. I shake out my legs and bend at the waist, stretching my spine, my hips. Beyond the fence is a ditch and a field. In the ditch is a man. He's lying down, dead or sleeping. A sound escapes me. The man stirs. He approaches the chain-link fence. He has one leg. He points to where another leg might once have been and then points beyond me, to the center of the Killing Fields, then back to himself. He holds out his hands. I reach in my pocket and find not much there. I give him what I have.

I complete the loop. The group at the last stop on the tour has grown even larger. I put my headphones on and press play. The audio guide tells me to approach a massive tree. The tourists make a circle around the tree, three people deep. They lean in, listening. Cameras flash. More people arrive. A plain wooden sign reads, "Killing tree against which executioners beat children." Everyone around me is interested, curious. The audio guide tells me bullets were expensive and so, to cut costs, the Khmer Rouge beat the heads of babies, toddlers, young children against the base of this tree and, if I look closely at the roots, I could see blood and bits of brain matter.

It is late morning. The full heat of the day has not yet arrived. The sky is seamless and swept clean of clouds. A handsome wooden bridge carries me over mass graves. The memorial stupa gleams golden in the sun. Inside the stupa are five thousand skulls. A dusty footpath, flowers sprouting at its sides, winds through gently sloping hills, below which thousands more bodies are buried. A man stands under an umbrella, selling ice cream and souvenirs. I keep looking around for signs of others experiencing catharsis and find only curiosity. I keep looking within for signs of self-knowledge and find only torpor. If there had once been a vague thought as to why I was here, it was gone now. If there had been a plan, it was disassembled. The birdsong goes on above. I fight the urge to lay down on the tranquil lawn with a novel and read. I am safe here. I understand curated spaces of pain. I understand the urge to guard the wound. Outside these walls are probing, restless streets. In here, I'm a living memory, suspended in static, cut off from the present moment. Outside, the moment arrives, arrives, arrives, is arriving.

The week before I left for Cambodia, I'd gone out in New York with a group of friends from the philosophy department. It had been someone's birthday. Jay had been there in the early hours but he'd gone home sensibly at midnight. The remaining group had wanted to go dancing and so did I. The first hours of the morning found me drunk in a corner booth leaning toward a handsome grad student. We had classes together. He was dull, possessed only one idea, which he rehearsed humorlessly in seminar paper after seminar paper. He was a consummate rule follower and an alcoholic. At school, he walked the hallways ravenous, always hungry, always thirsty. Once, on a break from class, a pizza materialized, and he ate it with real violence. He had a slippery way about him. He was too well-liked. I didn't trust him. He had a narcotic affability, it blurred edges. He touched my leg under the table. Our friends were somewhere dancing. His blue eyes watched me. I was submerged in an ambient lust. I did not respect him. He had an unmoored nature. I could not tell you if he had a single belief. This was what made him so good at following rules, not because he enjoyed deferring to authority, but because to go against authority, to rebel, would imply he had a stake in something. We pressed against each other and danced and then I stood outside the club alone, watching orange light glaze the Manhattan skyline. I took a cab home and crawled into bed next to my sleeping husband and son. They'd drifted off watching a nature documentary that was playing on a loop. Onscreen, a whale swam. It glided through vast darkness. I felt jealous of animal grace, elusive to humans, which comes not only from their movement but from their freedom from academic exercise. I remembered John Dewey's line: "The dog is never pedantic." Pedantism, he thought, came from the human tendency to sequester the past from the present in order to draw from it a model. But for animals, "The past absorbed into the present carries on; it presses forward."

Dancing with the handsome grad student, I'd felt outside of time or consequence. This felt familiar. And the hangover that followed the next morning furthered this hazy relationship to responsibility. I turned over Dewey's idea in my mind, that the separation from or absorption of the past dictated a direction in the present. I made Wolfgang's breakfast, rinsed his dishes, cleaned counters; time oozed in all directions like spilt and spreading milk. The night before, time had moved on a narrow track, racing from A to B, me to the man; us, under night's cover, pursuing pleasure. It was fun. And life at home was something else; it was cereal, soap scum, sunlight. I downloaded a dating app and idly scrolled through it, feeling nothing more than vague titillation. I deleted it an hour later. I didn't think of this, nor my night dancing too close to another man, as me being disloyal to my husband, nor did I think about the impact these actions could have on Wolfgang because I did not think that anything I did had an impact on anyone. I'd spent my life being told that I wasn't as real as other people.

I meet Chetra on the other side of the gates; he waves. He's only twenty-seven and so he was not alive during the genocide, but the other tuk tuk drivers are much older. They were children, teenagers, young men, when millions were marched out of the city, out to the village of Choeung Ek, handed spades and sticks and then were forced to dig their own graves. I'd noticed that Chetra kept to the edges of these groups of men and I'd thought this was his doing, that this was a sign of our similarity and that he, like me, felt separate from others. But now I see the men have not given him a choice. They hold themselves together in a locked chain of bodies, ringed tightly around the upturned box upon which they play cards, their backs turned out, a barrier protecting their interior space, keeping it accessible to them alone.

I say, "What did those men say about me?"

"When?"

"They pointed at me and said something to you when we got here?"

"They asked if you were my wife," says Chetra. "They thought you were one of us, but then when you went inside, they know you aren't."

On the drive back to Phnom Penh, I see a dog nursing a new puppy. The puppy is so young, his eyes aren't even open yet. The tuk tuk roars toward them and I wait for them to jump to safety as the other dogs and cats do. The mother makes it, but the puppy doesn't and Chetra's tire rolls over its neck. It yelps in pain, but only for a second. "He's done," says Chetra.

Chetra stops at a crossroads and turns around on his motorcycle to face me. He waves his menu of tourist attractions, then points to the picture of the National Museum of Cambodia.

"Do you want to go here?"

"Yes," I say. "But not today."

"Back here?" he asks, pointing to a skull.

"No," I say. "I'm sorry we ever went there."

"You're sorry?"

"I'm sorry I asked you to take me."

"That's how I make money," he says, and I see a flicker of annoyance cross his face. He turns away.

"I know," I say, and I almost apologize again, but stop.

"Where do you want to go?" he says, not looking at me. He's somewhere else, on the other side of the moment, he is wherever I'm not, wherever this conversation is over. "Just tell me where you want to go." And I can't tell him because I don't know and we're quiet for a moment, watching the day gather itself up over the fields surrounding us.

"Why did you come all this way alone?" asks Chetra. "What did you want to see?"

I tell Chetra about my now-abandoned project.

"You might have learned something else before."

"Before what?"

"Before Japanese companies leased the fields in Choeung Ek, landscaped it, put up gates, and turned it into a theme park. Tuol Sleng is the same, owned by the same company."

"Oh," I say, surprised and ashamed by my unresearched assumption that these places held meaning for Cambodians when perhaps they didn't or didn't in the way they once had before they became foreign-owned, for-profit theme parks for sorrow. Suddenly it hits me how obvious this should have been. In Tuol Sleng, there had been more space and emphasis dedicated to the story of a handsome, white American sailor who had been arrested and killed at S-21 after drifting into Cambodian waters than there had been to the thousands of Cambodian stories. They were all grouped together, a mass of faces blurred to facelessness in my memory. But the American man I could picture in my mind. He'd gotten his own separate place of honor in the museum. And I'd not even noticed this when I was there, so accustomed was I to things being adjusted to white comfort. Even Chetra, who I wanted to call my friend, knew how to mimic Western behavior to put people like me at ease.

"What did it look like before?" I ask him. "The Killing Fields?"

"It looked like a field. It wasn't labeled or separated from its surroundings."

Chetra's expression of disdain for these tourist spaces had demolished the way I'd been thinking about them just seconds earlier and a new set of ideas takes hold, but then I remember how young he is and how his perspective on these places came from a different set of experiences than those of the older men we saw outside the gates. Thinking of them, my thoughts shift again. I keep trying to take in too wide a view and every time I think I know what I'm seeing, new information destabilizes the landscape and the ground gives way below me. I had an idea that I could come here and be a neutral observer, which is what most people probably thought when they came here or went anywhere or maybe they didn't. Maybe everyone knew better. I feel stupid and lost. Chetra gives a rough shake of my shoulder. He raises an eyebrow. He has an idea.

"What?" I say.

"I know where we should go," he says.

"Where?"

"A secret," he says, and he gives his cart a decisive slap and off we go again.

Chetra drives us to the edge of the muddy Mekong. I trace a line with my fingernail through the dust on my skin. We stop at a ferry landing and some men help Chetra pull his motorcycle and cart safely onto the small boat and we take it across the Mekong to an island. *Koh Dach*, Chetra tells me. *Silk Island*. We stop at a cart for grilled meat and soda and eat it in the grass near a grove of mulberry trees. After, we walk below their branches and Chetra points out silkworm pods, speckling the leaves like mossy white berries. We ride out along a rough country road that winds through a village. I see a white man standing in the doorway of a low-slung shack. The man watches me intently as we go by in the tuk tuk. Chetra notices and tells me white men come and give him lots of money to help them to experience "the ladies."

We crest a hill and there is a beach leading up to the Mekong. We walk across the smooth sand to the edge of the river. Wooden planks lead from the shore out to small huts that hover above the water.

"The secret beach," says Chetra with exaggerated reverence. Then he laughs. He's teasing me the way he'd teased his cousin. It's clear that this is not a "secret" beach, but another tourist spot. But this one sells Cambodia's beauty. A woman approaches and I pay her an entrance fee. Children gather around to show me their wares, bracelets made of woven string.

The slim plank that stretches out across the water wobbles. I can't balance on it by myself. I can't get across. Chetra has run ahead but turns back when he sees me stranded on the beach.

"What's the best way?" he says. He does not grab me. He does not immobilize my arms, thinking this is helpful. He offers me options, points to his elbow then his shoulder and then he extends his hand. "How do you want to go?"

I put my hand on his shoulder and we inch along the wooden plank this way, out toward the hut. This gesture toward my autonomy, this willingness to ask me what I need, it's so rare. When Chetra extends this grace to me, my first thought is to consult the list, pinned to the wall in my neutral room, of all the times people have denied me theirs. But then a second thought comes, which is that I'm walking on a wooden plank out over the surface of the Mekong River and that I ought to pay attention. A bit of resistance in me eases and I feel a little lighter, more surefooted, and, with Chetra's help, I make it across.

The hut is made of bamboo and the roof of thatched palm fronds. It sits a foot above the water. I watch the horizon line until my eyes ache from the sun and then I lie back and stare at patterns in fronds above. I close my eyes and miss Andrew and Wolfgang. I hold the image of Wolfgang behind my eyelids and hear a child's voice and then a splash. I sit up and look out over the river. A flash of colorful fabric pulls my eyes up into the sky. There are three young boys, one in the water swimming, the others in their boat, flying a kite. They float down the river. Later, they dive off their boat and swim up to us and sell us mangoes. I buy one for me and one for Chetra and we eat. A long time passes.

"So why did you take me here?" I ask.

"It's far so you'll have to pay me more," he says, his eyes gleaming and good-natured. "I really thought you'd like it here," he adds. And I do.

Two kids swim up to the edge of the hut and talk to Chetra. They want to know if he'll come swim with them, but he tells them no. They ask again, reaching up into the hut to pull on his legs, trying to move him into the water. He shoves the kids aside and they fall back into the water laughing. They splash us and swim on.

"I don't like the water," Chetra says in a tone that's shifted so quickly that I whip my head to look at him and knock myself off-balance. He's scared. "Ghosts are in the water," he says. "Can I show you something?" I nod and he opens Facebook on his phone and shows me photos of his brother's four children.

"Where is your brother?" I ask.

"You don't remember?" he says.

"Remember?"

"He died."

"I'm sorry," I say.

"Motorcycle accident. Two years ago."

"I'm so sorry," I say.

"My cousin told you at the restaurant, remember? After we saw the man who died in the street?"

I think for a second and then my stomach drops. I remember the cousin saying something I didn't understand, and I remember that instead of asking for clarity, I'd laughed in response. I remember Chetra's hurt expression and the way he'd snatched the nursing cat from my lap in the restaurant, the sweat that had poured from his face when he'd returned.

"Oh, I didn't realize . . ." I begin, but he puts up his hand to stop me from saying more. I don't know what to say to him now other than *I'm sorry, I'm so sorry.*

"I look out here and think of my brother and wonder, why is it he had to die? Time goes by and I miss him more, not less."

I feel myself sinking down into exhaustion. The day has drained me. I feel awful for our misunderstanding and my role in it. I am also exhausted by the effort it takes to understand Chetra, to listen so closely. Language is a barrier, perspective is another, experience a third. But this is not the real issue. The real issue is that I am trying to bridge a gap between us while simultaneously holding myself apart. These conflicting desires in me wear me out and urge me to forget the effort. I have fantasies of sleep, of air-conditioning, of baths, of bad television I could stream on my laptop, of room service. I do not want to be in this hut on the river listening to the story of someone else's loss. For a moment I wonder if I had understood his cousin but had chosen to close my mind against her words. I did not want to take on his suffering.

This feeling I had that my disability kept me on life's sidelines was painful but could also be a helpful delusion. If I'm outside the boundaries of the real world, then I'm incapable of influencing it or inflicting harm. But I was in the world, and I could cause harm, as much as anyone else, more even, if I pretended I was exempt.

The light reflected off the water draws shapes on Chetra's face and I watch his face, which changes, looking new with every passing wave, with every change in light. We turn back to the beach. He holds my hand, which holds his shoulder, and he helps me across the narrow wooden planks, back to safety.

On our way back to the ferry, I see the same white man standing in the doorway of the hut. He looks at me. His eyes are blue. He'd come to take an experience for himself alone and so had I. He lifts his hand to his head. A salute.

I leave Phnom Penh by bus and travel to Siem Reap. There I see temples until I am sick of temples. In Battambang, I watch bats rise from a cave by the thousands. Then I come back to Phnom Penh. Chetra had messaged me on Facebook, asking when I'd return. I tell him my time and date and he says he'll meet me at the bus station and take me back to my hotel. I'll stay in Phnom Penh a few more days and then will fly home. I look for Chetra's face in the crowd of tuk tuk drivers waiting at the station, but he's not there. I walk to the edges of the crowd, look for where he might be waiting, but he's not there. Another driver takes my bag and I let him.

On my last full day in Cambodia, I hear a knock on my hotel door. I leave for the airport early the next morning. A woman is standing outside my door. She says something in Khmer and I tell her I don't understand and I don't know what she wants. She repeats a phrase I can't comprehend. She then walks right past me, into my room. She picks up the hotel phone and calls someone. She hands me the phone. On the other line, a concierge tells me that a free massage comes with my stay. She is there to massage me. I hang up the phone and look at the woman. She motions toward the bed. I shake my head no and tell her it's OK, I don't need a massage, but she takes my arm hard and leads me to the bed. She swipes the air in front of me as if that might simply erase my body and with it our misunderstandings. I ask her if she's telling me to take off my clothes. I don't know if it is customary to take off your clothes if a Cambodian woman insists on massaging you. I take off my clothes and lie facedown on the bed and she slaps a bit at my back. This goes on for a while. Then she stops.

She stands somewhere in the room. My face is smashed into a pillow. I turn to peek at her. She's looking down at me from the end of the bed.

I ask her, "Should I turn over?"

She hushes me with a curt word.

I go back into my pillow. I do not dare look up again. I wait. Then I hear her voice. She's whispering something from some corner of the room. She keeps whispering. I know only enough to know that she's not talking to me. Then I hear the toilet flush. She's in my bathroom whispering. Minutes slip by, then more, more. I am confused and then afraid.

No one knows where I am. Absolutely no one anywhere in the world knows where I am. I had only told Andrew I was going to Cambodia. I'd not even told him which city I'd be in, let alone which hotel I'd booked. I'd just gotten onto a plane and slipped out into the world, alone, wanting to disappear and maybe this woman was here to help me or force me to disappear and maybe she was talking to someone who would show up and disappear me and was that what I wanted, yes, but also, no, how awful, how lonely, how alone I'd made myself, how far I'd kept everyone from me, how hurt and isolated I was and just then the woman comes out of the bathroom and stands at my sink, washing her hands for a long time, warming them again and again in the hot water and she comes to my side and touches me gently and I sit up and she covers me with the blankets and then gets into the bed behind me with her back against the headboard and motions for me to put my head in her lap and I do and she touches my temples and my scalp and runs her fingers through my hair and massages my head while I weep. Pain dissolves under her touch. But I loved my pain, I protected it, I kept my wounds clean. I mourned its absence. I relied on it to muffle the drumbeat of junk in my head.

I've found so much solace in engaging in things that were fixed—dead philosophers and their theories. Art on the wall. Stone statues. In these relationships, I was the dynamic thing. I named, analyzed, judged, moved against the unmoving. I diagnosed and demarcated from within my neutral room, which, I begin to see—and of course this is obvious, and something I should have known much earlier in life, and I would have seen this sooner had I not been so busy building my little cocoon of self-regard—was not so neutral.

I easily could spend my life keeping Andrew, Wolfgang, and everyone else at a fixed distance. That was a choice I could make.

My father saw himself as a great protagonist, and therefore everything that happened, happened to him. I was less grandiose in my thinking but no less pernicious. I felt I was outside of the story altogether and thus free to act without consequence or awareness of my power to hurt others. We were Janus, one head looking in two directions.

In the afternoon, I go outside, hoping Chetra will be there, but he's not. I ask the clerk at the counter if he knows where Chetra might be and he shrugs.

"How was the restaurant I suggested? Very good, right?" asks the clerk.

I tell him I never made it and he looks disappointed.

"Tonight," he says, "you go. This is your place! Best place for you."

I go back outside. I get to the end of the alley and behind me I hear a voice I know.

"Where are you going? I'll drive you."

I turn and Chetra is there. And I am so happy to see him, my friend. It breaks me free of some dark space, a space I'll never return to.

I tell him again about the concierge's recommended restaurant and he rolls his eyes.

"His sister owns it," he says, laughing.

I laugh, too, and then, a shaky deep breath escapes me. I'd felt very sad at the thought of leaving Cambodia without seeing Chetra again, without the chance to say goodbye.

"OK," he says, "I'll come with you."

"I can get there."

"Not alone. Not walking. You don't know how to cross the street," he says and he is right. There are no crosswalks or stoplights where we are and I have no idea how to stem the flood of motorcycles long enough to get from one median to the other. Chetra and I walk side by side past the alley, down the block, and, when we arrive at the bursting street, he simply enters it, arms waving. A motorcycle slows and swerves around him, as do the others that follow. Chetra's body, a rock the tide rushes around. I stand behind him, safe in his wake, and we walk like this across the street.

The concierge's recommended restaurant is American-owned and serves Chinese food. Inside are only Australians. It feels hokey and I'm embarrassed to bring Chetra here, but I also want dumplings. We order. I look down at the napkin I've just used to dry my face of sweat and it is brown with dust and I laugh and show Chetra. The coat of dust on my skin, my black fingernails, the gritty feeling between my teeth—it was so familiar that I didn't notice it anymore. Chetra shows me the ruddy smears on his own white cloth. There is no way to go into the city and come out clean.

Chetra quizzes me on the Khmer he's taught me. I pass to his satisfaction. He teaches me more. Chetra pokes the gelatinous skin of his pork bun and says, "I think this is not food." He pokes it again and I laugh. He pulls it apart with a fork and looks inside, then comically turns away. "This is not food." Then he takes a bite and says, "This is so good."

And it is so good. The concierge was right. This was my place.

After dinner, Chetra says, "Where to now?"

"Where do you actually like to go? Not a place just to take tourists, but a real place."

"She's come to see the real Cambodia!" he says, laughing at me. "Take me to the authenticity, she says!"

"Take me to the authenticity," I say, and he laughs again.

"OK," he says. "I know where it is," and he starts his motorcycle and off we go into the cooling evening, through the city, over a bridge, and then I see lights, bright and spinning. "Fantasy World," says Chetra, and I don't know if it is the name of where we are or a name for something else.

The theme park grounds are packed with people. Families wait in long lines, their children on their shoulders. Kids whiz by us on bikes. Speakers hanging from their handlebars bleat competing melodies. Chetra buys us French fries. I get us two Cokes. He points to one ride and then another. There are so many, and they are lined up so close that when I look up, I see a roving canopy, the arms of mechanical creatures flinging, spinning, twirling. Everywhere, heat and screams, worm-eaten leaves, animal excrement, delight, the decaying day. The sun sets, uprooting resting hues before disappearing, leaving the moon to settle the sky's riot of copper colors.

First, we ride a roller coaster, then the Ferris wheel.

9

Above the Middle Range

My flight back to Brooklyn from Cambodia includes a ten-hour layover in Taipei. I sleep on a bench in the airport and dream of Wolfgang, who runs into my arms when I finally land in New York. I'm so happy to see him, my Little Shadow.

I've missed a parent-teacher conference, Andrew tells me, in which Wolfgang's teacher expressed concern about Wolfgang's fixation on the past.

In class his face would change, she told Andrew. His eyes would dim like two dark clouds, and he would become very quiet, his self somewhere lost, his being boarded up. And if pushed to speak, he'd only say, *I'm having that feeling again where I'm sad about the past*.

He'd lost all interest in recess, choosing instead to stay inside and talk with any adult who was willing—teachers, janitors, the front office secretaries—while they had their lunch. They'd ask him how he felt, and he'd say he was in a panic, knowing the past was gone.

This had been going on for a while.

Later that week, Wolfgang and I go on a walk around our Brooklyn neighborhood, and he looks up at me and says, "The day we got Sose was the best day of my life." Sose is our cat.

"That's so sweet," I say.

"And I can never get that day back again," he says, and he starts to cry.

His sadness is either linked to or is simply growing alongside another worry.

Wolfgang, who turned six in the fall, is certain everything is alive and in need of more understanding than it is receiving. Living or not, he calls everything his friend—sticks, flowers, every animal he sees, every drawing he makes, and also his old clothes, which he will not relinquish as he grows out of them, even his too-small socks.

"I want to keep them for the memories," he'd say. "And I don't want them to be lonely."

Iris Murdoch had written in her diaries about a similar intense animism she felt for all things—"cats, buses, stones"—in the natural and man-made world. "My friends, my friends, I say to the teacups and spoons," she wrote.

Wolfgang also feels awful for Food. Chewing is agony. Eating hurts Food's feelings.

"I tell myself," Wolfgang says, "to get through it," meaning to get through the act of eating, "that I'm just teaching Food a lesson in disappointment, early in life."

It is crucial, he explains, to learn lessons in disappointment early to be prepared for bigger disappointments later. Reflecting on this is the only way he can manage to eat Food. He believes, although it pains him to do so, that inflicting disappointment on Food now would help Food later to become more resilient.

"Interesting," I say. "Wolfgang, have you learned lessons in disappointment early in life?"

"Yes," he says, "I've been disappointed before, I will say that."

"Can you think of any examples?"

"Sometimes you tell me we're going to a birthday party and I get so excited and then it turns out to be just all adults having dinner at a restaurant."

"Is this normal?" I ask Andrew later that night.

"His teachers don't think it is the most normal, no."

"What does it mean? Where's it coming from?"

I try to make everyone watch tennis with me. Bobby relents and we spend hours on the couch, half-watching the matches but mostly talking about his adventures on Tinder. His divorce had gone through. He missed the cats he'd shared with his wife, which she kept in California. He looked at pictures of the cats and cried sometimes. He was doing great on Tinder. He'd met an artist named Jess who he felt might be someone he could love.

Jay came over to watch the women's matches. He liked Serena Williams, who had recently won the Australian Open while pregnant. Roger Federer had won the men's draw of the same Australian Open. He'd been off the tour for months, recovering from knee surgery, and was now, in his mid-thirties, playing some of the best tennis of his career. It was described in the press as a divine resurrection. It was the second coming of the true messiah of tennis, said John McEnroe. It was a new age: "Welcome to A.R. 35." He was, claimed another journalist, "a permanent miracle." A writer in *The New Yorker* wrote that Federer embodied a Platonic ideal: "It is like rooting for the truth."

When I was in college, I drank at a bar that always had tennis on. I'd watch without understanding. All I could see in front of me were people running and whacking a ball at each other until one stopped or gave up or made a mistake. If there was more to it, I couldn't see it. The bartender tried to explain things to me, but as I listened, I'd have the feeling that we weren't watching the same events unfold. I would listen to his definition of, say, "wrong-footing" someone, but I didn't understand the concept because I couldn't *see* it when it happened on the screen. He'd say, *There, look, he wrong-footed him*, but all I'd see is a ball whizzing over a net and off the court.

I did understand that a ball could be hit with a stroke referred to as either a forehand or a backhand, but I didn't get what it meant to "hit to" the opponent's forehand or backhand. I was baffled by the idea that a good player could "read" their opponent's serve. The serves were so fast and all looked the same. The level of fine discernment needed to truly understand the game seemed beyond my abilities and it upset me to be so left out of both a verbal and visual language.

But when I saw Roger Federer play for the first time, all those years ago, a strange thing happened. My perception was sharpened, briefly, allowing me a heightened moment of noticing. Federer's backhand was beautiful and I didn't need anyone to explain that to me. I could see it. He opened a door and let me in. The more I watched him play, the more I learned. He became my teacher, my translator, and so I became devoted to him. I sought out all his matches. I began to hate his rivals. I moved forward, learning more nuanced aspects of his game. I watched videos of his footwork. I listened to people analyze his kick serve. I found out what a kick serve was. I felt an urgency to be in his presence, a summoning. He inspired nearly deranged reactions from his fans. I was fascinated by the awe I saw on their faces when the camera panned to the stands.

I'd read in the *New York Times* that tickets to the Australian Open final, where he'd played his old rival Rafa Nadal, had sold for close to $20,000. "It's priceless to me," a ticket holder explained. "It's pathological," said another fan, explaining his need. They sought to be close to him, not separated by a screen. What did anyone gain, specifically, from proximity to genius? I wanted the answer. I wanted to know what these other people knew, feel what they felt. I wanted to know if these ecstatic visions were available to me.

Bobby and I are invited to a party held at a bar. I don't know anyone there. I'm thrilled when I see tennis on one of the big-screen TVs and I sit and watch. Next to me on a stool sits a stranger who is also watching tennis with my same serious interest. We start to talk about the recent Australian Open. He teases me when I speak reverentially about Roger Federer, whom he deems boring and a too-obvious favorite. He, a true original, prefers Nadal.

We keep arguing about tennis. He's funny and smart, if wrong about Roger Federer. He gives me his card before Bobby and I leave for home. It turns out that he is an editor at *GQ*. The next day I email him and say something like *Let me write about tennis* and he doesn't respond.

I keep on it. I feel a compulsion to be near Roger Federer and sense this *GQ* editor is my golden ticket. I send him my articles from Sundance and the article about the Hannah Arendt documentary. He likes them and eventually, when I ask to write about tennis again, says something like, *Sure, who cares, why not.*

I am issued a press pass to attend the Indian Wells Masters tournament near Palm Springs. I am alone in bed when this news comes through. The full scope of my inexperience hits me all at once. Not knowing what else to do, I google "tennis" and start to read. I stop. I slide off the bed and onto the floor and stay there, my face down, inhaling the chemical cleaner in the carpet.

I walk out to the living room to share my news. Andrew and Bobby and Wolfgang are stretched out on the couch watching a movie, laughing. They do not need me, I think, and so it would be OK for me to go away again.

Someone sings a sad song outside my window. It pulls me half out of sleep and I listen, unsure for a moment where I am. Brooklyn on a Sunday morning, people singing on their way to church—but no, it is not Sunday, it is either Tuesday or Wednesday, and I am alone in bed, out in the desert, far from home.

I look at the clock. An alarm on the other side of the country would be waking Wolfgang for school. It is too early in California, but my sleep cycle is synched up with the rhythm of home. Andrew would be saying, *Wolfie, Wolfie, it's time to wake up*, over and over. I stare at the clock. The sad singer cuts into my visions of home. I listen and watch soft light spill across the dark stucco ceiling above me. The song is familiar, an old soul song. I am certain now that its melody will be etched on me, looping in my head for the rest of the day, but then I fall back asleep, and when I wake again the singing outside my window has stopped and the song is wiped from my memory and the more I grasp at it, the more I try to remember just one single word or note, the further it slips until I have only impressionistic glimpses of a memory not an hour old: the prior coolness of the early morning, the rustle of sheets, the hesitant glow of dawn, a voice echoing in the desert somewhere beyond me.

Day four of my "coverage" of the Indian Wells Masters tournament and my strategies have not evolved past moving from one hiding spot to another. I hide in the bathroom. I hide in the hallways that snake below the stands. I hide in the far corner of the stadium's designated press seating. I hide in the back of press conference rooms while the real journalists ask tennis players questions. There isn't much else for me to do other than hide and sit and watch. I could write something, but what? I don't know. My observations are few: there are winners and losers and after one match comes another.

The real journalists write so much they don't have time to watch what they're writing about. Muted TV screens glow above every press desk, displaying the same match unfolding directly outside the press office. To see it live, all one would have to do is turn their head to the window. These journalists don't do that and yet they know all the scores, but not just that, they know how the players have played and exactly what they've done to make themselves succeed or fail. They murmur across desk dividers in their secret language. *Takes time away*, one says. Another says, *Such big cuts.* Later, *To the body. Jammed her*, and then, *Out wide, cross-court, one, two: point, point, point, game.* I don't understand my surroundings nor the people within them, and because of this I am doubly isolated: first by being ignorant and second by pretending I'm not.

At the end of each day, after speaking to no one and learning nothing, I drive a long drive alone, along dark desert highways, through mountains I can't see in the darkness, to my hotel, where I sleep alone, then wake, and return to do the same the next day.

The desk in the press room next to mine is occupied by a young Australian journalist. All day he eats salads and types furiously, filing six or seven articles at a time, it seems. I catch him looking up from laptop screen to TV screen and then briefly he turns to watch through the glass the actual match. This is an opening and I gather the courage to ask if he's hoping one player will win over the other and he says he doesn't care when a player he likes loses. I stare at him, hungry for conversation.

"I used to care," he continues. He eats a red pepper sliced thin as a fingernail. He keeps extra vegetables in plastic containers stacked on his desk.

"It's a job, though," he says. "There will always be another tournament. They'll play again."

My only friend is Judy. Judy runs the main press information desk. It's her job to assign workspaces and lockers to the journalists. Judy welcomes me every morning by saying my name.

"Don't forget your coffee, Chloé," Judy says and I say back, "I never do, Judy."

She hands me a player interview request form and I nod my head like interviewing a player is something I'll definitely do, just as soon as I get settled in. In the afternoon, she brings us the schedule for the next day's matches, which everyone calls the OOP, the abbreviation for "order of play." When she stops at my desk, she says, "Almost time for the afternoon cup, Chloé?" and I'll say, "You know what, Judy. I think it might be just that time."

There's a certain Lady of Palm Springs look and Judy's got it—dyed blonde hair, big diamond rings, perfect nails, perfect makeup. Judy has fewer wrinkles than I do although she's likely thirty years my senior. She tells me that she volunteers every year at the tournament just to keep herself from "rotting away in her condo, ha ha!"

Each morning, I greet Judy, pour myself a cup of coffee, and walk past the rows of desks in the press office, aching to make eye contact with someone who might recognize me from the previous day and who will acknowledge me with a nod. Every time I pass by the rows, I stare at the focused faces of the actual journalists. If they so much as lift their gaze for a moment, I'll be ready with my acknowledging nod. The TVs blaze above their heads. Everyone is very busy. A sports newsroom is a place where if I said, *I'm having a heart attack, please help me*, the response might be, *Quiet, please, I'm on deadline*. When I get to my desk, my TV screen is dark and I can't get it to turn on. This is a new problem. I run my hand along the screen's cold, smooth edges. There are no buttons. Judy, on a surveillance stroll, passes my desk and sees me struggling. She pats my shoulder gently and says, "Watch this, Chloé," and then she touches the center of the screen itself. Instantly it comes to life, bathing Judy's face in its electric glow. I thank her profusely and she says, "You're very welcome, Chloé," and then she says, "Have a great day, Chloé." I want to get down on my knees in praise of Judy.

At the end of the day, I hear the young Australian journalist talk to the cool French journalist who sits on the other side of him about having dinner. My stomach growls. To her he whispers, "I'm terrible, I snuck some bread for breakfast and bread for lunch." I clean up my desk as they clean up theirs. A debate about where to eat slows them down, so I pretend an important email has come in and type random sentences in an empty Word document until they start winding up their laptop cords and then I wind mine and then they put their MacBooks in their sleeves and I dump mine into my backpack and when they are ready to go, I'm ready, too, but they walk out, forgetting to invite me to dinner.

I drive to my hotel with the windows down and I breathe in the air, which is deeply perfumed, something I didn't expect in the desert. I was sure it would smell hot and dry, which is to say I thought it would not smell like anything, certainly not gardenias and marigolds.

I'm grateful for the cool darkness of the desert at night. I pass my hotel once, double back, and then pass it again, just so I can prolong the feeling of being hidden away in the dark car, in motion, moving forward not in failure.

I forget I'm not supposed to cheer in the press box. Lord, do I forget. The very serious, very busy journalists glare. I whistle and gasp. Some move to sit further from me. I expose my inexperience again and again and I must go hide in the bathroom, looking in the mirror, smoothing my hair. When I return to the press box, a *Times* reporter I recognize moves to sit next to me. He introduces himself and I shake his hand. He asks me if I'm reporting on anything specific and I drop my phone to avoid answering. When I sit up, I notice some people in the crowd are looking back at me. A few more heads turn. Play on the court below has stopped. The chair umpire points to me. A feeling of illness pulses through me. I've done something wrong and now all are ready to jointly expel me. The *Times* reporter spots the problem. He reaches for my hand. Somehow, in my fumbling, I'd turned on my cell phone's flashlight and it was facing out, the beam reaching down onto the court, disrupting play. Tennis is a sensitive sport. The slightest distractions are rooted out. The *Times* reporter takes my phone from me. The flashlight blinks off. He sets the phone back into my open hands. Play resumes.

I can prolong the three-minute conversation window I have with the young Australian if I, too, eat vegetables.

"They're so crunchy," he says, smiling at me as I pass gravely by, my salad extended in front of me like a head on a platter. I say, "Yes, crunchy," and force a weary grin and then I say, "Good lettuce." I shrug. He looks at me, at my pudgy, wobbling body, and says, "Don't worry, when no one is looking I go around the corner and stuff my face with doughnuts." And I say, "I bought the salad because of you and your little boxes of vegetables," and he says, "I'm a great influence," and he winks and now I have a friend and the next day when I return, I touch his shoulder and say, "Good morning" and he looks at me like he might know me from somewhere.

I brought with me a collection of travel essays by the writer Geoff Dyer and each night in the hotel I read the essays and pretend I am him. I read a little Laurie Lee and pretend to be him, too. I like these writers and the others of their ilk. These solitary men who wandered the earth, trusting their eye, their intelligence, their strong bodies, to carry them safely across vast space. These independent men in pursuit of the upper limits of human freedom.

Dyer's expertise is rarely if ever the point of his essays, which often end up as odes to strange accidents, mistakes, failures, sex adventures with strangers. I try on his curiosity, his bravery, his humor. The next morning, I walk the grounds of the Indian Wells Tennis Garden believing I am Geoff Dyer. I greet people as I imagine he would greet them, I get my coffee the way he would, I sit at my desk and open my computer and begin to write like I have a reason to be where I am, doing what I'm doing. I begin to write a draft of a story about the crowds surrounding Federer's practice. Then I start writing a draft of another story and another. I make a choice to trust myself, to trust my ability to observe and to write down these observations. I begin to believe it is possible to simply dismiss the concept of defeat.

Federer has his first press conference and I find that, even in my Geoff Dyer mental costume, I lack the courage to ask Federer a question. The mere thought of raising my hand and speaking out loud in this room of serious journalists fills me with dread. I want so badly to ask Federer a question but want more to keep eyes off me and my many mistakes. I sit quietly, steeping in my humiliation over a merely imagined moment.

Out of the corner of my eye, I see a tall figure with white hair pass my desk and I imagine for a moment that it is Geoff Dyer strolling by, checking on me, appraising my progress. Later I climb to the top of the stadium and look back over its edge down to the pools of people outside, drifting around the grounds, and I think I see Dyer; his white hair bobs above the crowd for a moment before sinking back into the masses and disappearing from my sight. For a moment, I catch myself, aware of what I'm really doing. Reading Geoff Dyer, I picture my father. The two look alike. Same blue eyes. One is the man the other wished he'd become. Reading a book, writing at my desk, looking into the crowds, my father is nowhere and everywhere.

Driving back to the hotel that night, I'm caught in a brief sandstorm. The view through my rental car's windshield goes cloudy. I drive slowly, trying to decipher the markers of meaning—the road, the yellow dividing line, the presence of other cars—I need to keep myself safe. My mind and senses are sharpened and, while this is a moment of real danger, I feel that rush of pleasure that attends the experience of one's heightened faculties.

Back at the hotel, I watch the sandstorm safely from behind my closed window. Kant would name the feeling I'm having as an experience of the Sublime, which he describes in one of the more famous passages from Western thought:

> *Consider bold, overhanging and, as it were, threatening rocks, thunderclouds piling up in the sky and moving about accompanied by lightning and thunderclaps, volcanoes with all their destructive power, hurricanes with all the devastation they leave behind, the boundless ocean heaved up, the high waterfall of a mighty river,*

and so on. Compared to the might of any of these, our ability to resist becomes an insignificant trifle. Yet the sight of them becomes all the more attractive the more fearful it is, provided we are in a safe place. And we like to call these objects sublime because they raise the soul's fortitude above its usual middle range and allow us to discover in ourselves an ability to resist which is of a quite different kind, and which gives us the courage to believe that we could be a match for nature's seeming omnipotence.

I drift off to sleep as if in the embrace of such courage.

The next morning, I watch matches, check my email, hide in the back row of press conferences. Walking to lunch, I pass a very tall tennis player on his way to the practice courts. I recognize him and as he sees me he stops as if he recognizes me, too, although that's impossible. He's confused by something and stares at me, looking up and down the brief length of my body. I'm walking toward him and he starts to mimic my swayed walk. He nudges the person next to him, likely his practice partner, and says, too loudly, "What the fuck is that?" as he passes me.

And I think, *There it is!*

And I feel relief.

A year ago, for Christmas, Andrew and Wolfgang and I tried to take a trip to Skaneateles, a small town on a lake, five hours north of our home in Brooklyn. We were nearly there when a snow squall descended. It dropped like a pearled curtain, obscuring the world beyond our headlights. Andrew slowed the car. Wolfgang remained unconcerned. He nodded his head, oblivious, smiling, lost in some other scene unfolding on a screen he held in his lap.

A few flurries in the wind, winter imprinting peacefully on the pastoral scene. Hills in the distance, pocked by the shadows of cattle, grew a silvery film. We went a mile, another mile, and another, and then, all at once, an invisible seam ripped open and the world flipped white. The snow squall became a shroud, a winding-sheet of white surrounding us.

I focused on the space between the left curve of our car and the caution line that separated us from oncoming traffic. Our safety depended on the maintenance of that space between that yellow line; on the other side was speeding metal. I stared there, concentrating, willing this gap to hold.

Two silver dimes appeared, hovering in the white. Andrew's hands paled on the wheel. When assessing our threats, we'd forgotten about the reality of other cars; we could be careful but we could not keep someone from slamming into us. Wolfgang sang along to a song I couldn't hear. I felt my feelings rise above a middle range and felt a brief, sublime courage surge in me. I felt my own mind could control our car and the one that approached. I tried to determine how far it was from us, how much time we had to brace ourselves, but the car was already upon us, whooshing up, headlights blaring, and for a terrible moment it was closer than anything, leaning near, but then the moment passed, and I watched in the rearview as the car, reddened in our taillight, receded.

The snow squall stopped as abruptly as it began, and we arrived in Skaneateles unscathed. Mansions loomed, mostly Greek and Italianate revivals guarded by ornate iron gates. The town was founded and seemingly frozen in the 1800s. We drove along the main drag, a single street of retail shops jammed into Federalist-style buildings, stretching itself out along the head of the lake.

Our bed-and-breakfast sat on a hill. Its sloped roof was iced in perfectly even layers of pristine snow. Candlelight flickered in the windows. I'd booked our stay after first seeing the hotel on a friend's social media account. She'd taken her family there in the summer. In their pictures, her two children posed in the middle of a lush garden.

These kids were, according to pictures, living a life of nonstop wonder. They were constantly getting lost in corn mazes, dancing around campfires, or spelling out words with shells on a beach. These kids hiked and swam and loved their dog. Occasionally my friend's hot husband could be seen in the background splitting wood with an axe. I would have ordered this family from a catalogue. The mother, my friend, looked as thin and stylish as she'd been at nineteen. There was an apparent alignment in all the elements of her life. She made it look easy.

The B&B's rough stucco walls were painted an antique yellow, a sun-faded yellow, and I wondered how the effect was manufactured. I imagined someone scrubbing the exterior with a wire-bristled brush to age the paint. The warm candlelight glowing through the glass waved, flashed, reset; it was fake, electric, flickering on a timed interval. I frowned. Andrew caught it and nudged me. I smiled my most reassuring smile to let him know, *I will not ruin this good time. I will not muddy this family memory.*

The storefronts in downtown Skaneateles were covered in Christmas decorations. Wolfgang pressed his nose against the candy shop's glass panes, which were smudged and stained with grease from fingers. The window's edges were ringed with thick fake frost. Inside, a shop clerk handed out free treats to children. He smiled and motioned for Wolfgang to come and I felt my hand, a somnambulist's hand, reach out involuntarily and grip his shoulder, stopping him.

"It's OK, you can go in," said Andrew, and I released Wolfgang's shoulder and nodded.

"Of course," I said. "Go ahead." But Wolfgang watched me, leaving my side reluctantly.

The clerk placed something in Wolfgang's palm, curling his fist shut over a treasure, and the clerk said something to Wolfgang and Wolfgang smiled and the clerk touched Wolfgang's hair gently. I did not like this. It felt staged and suspicious. Wolfgang beamed up at this man with wide eyes, then saw me waving him back to us and away from the scene, and he dutifully returned to my side, stifling a thrill.

Andrew and I looked down at him as he unfurled his fingers, one by one, drawing out the gesture for the sake of the big reveal and: it was a gumdrop. He bit into it, yowled, spit it out. I tasted the remainder. It was black licorice.

We walked around Skaneateles. I could not keep from ridiculing the endless quaintness. At the end of the street was a park with a gazebo. It was on the edge of the gray lake. We walked around the park until we were tired and ready to go back to the B&B. We were now far away from where we'd parked and so Andrew left us on a bench and sprinted off to retrieve the car.

Wolfgang and I watched a family take pictures of themselves inside the gazebo. They were us in an alternate timeline, a man and woman my and Andrew's age, better dressed for the weather, taller, smiling. At their hip was a boy, burly and unconcerned. They left the gazebo and walked over to admire a Christmas display in the window of a real estate office. I watched this family and imagined a version of myself that could just enjoy this nice town, this snowy afternoon on a lake, the day before Christmas, and then I felt very tired, and I leaned back against the park bench and I closed my eyes and when I opened them again the family was walking toward me, trying to get my attention. The mother was staring at me, telegraphing concern. She opened her mouth, but I could not hear what she said. I turned to my side and Wolfgang was gone.

He was running away from me, down the pier, toward a Christmas tree on the end of a dock stretched out over the freezing lake. Signs warned not to walk out on the pier. Wolfgang's arms were stretched out in front of him and I realized he meant to climb the tree.

"Wolfgang!" I yelled.

I tried to chase him, but he was fast. I lost my footing in the snow and fell.

Imagine your child is running out into the street and you can't run out and protect him, my doctor had said. Wolfgang was sliding in his rubber boots down the pier. *Your body could endanger his.*

I shouted, "Stop. Wolfgang, stop."

This pregnancy is dangerous. For you. For your child. My doctor had offered to advocate to the state for me to receive a late-term abortion. He'd thought my hips would separate in my third trimester and I'd lose the ability to walk, perhaps permanently. He'd believed my frame would blunt fetal growth. *What if he is born like you?* What if I fell in the snow while chasing him, failing to save him from the danger of a rickety pier stretched out over a frozen lake?

The father from this other family ran after Wolfgang. He caught him at the end of the pier. I heard him say something in a stern voice and then I heard Wolfgang howl. He'd just wanted a closer look at the pretty tree. The father picked up my son by the waist and deposited him at my side. There was snow in my sneakers and my socks were soaked.

"Thank you," I said, holding Wolfgang in my arms.

"You need to be more careful," he said. "He could have fallen in the water. Is there someone else here with you?"

I turned my back on this father, feeling shame.

But also I felt relief. *There it is.*

There was the judgment I'd spent the whole day waiting for. I knew someone would identify me as inferior. I only had to look for it. *There it is.*

I looked down at Wolfgang. He saw plainly my disgust at this man who'd tried to help us. Wolfgang took my hand, watching me carefully. I could see my anger seeping into him, killing off the thought that a stranger could be kind. When I was pregnant, the perceived dangers that accompanied my body were all anyone talked to me about. No one was worried enough about how my mind could limit him.

On our last day in Skaneateles, we saw a Santa on the street. I said to Wolfgang, "Do you want to take a picture with Santa?" and Wolfgang shook his head no.

The Santa called Wolfgang over, but he refused to go. The Santa waved again and Andrew and I urged Wolfgang forward. The Santa crouched down to receive him and said, "OK, little boy, what do you want me to bring you for Christmas?"

And Wolfgang said to Santa, "You don't have to do this. I know you're just a regular guy in a suit." The Santa had his arm around Wolfgang but he broke free and ran back to my side. I looked down at him, my gentle, sensitive son who watched me so closely and drank up my moods, who looks to me for the truth. I didn't want to ruin things for him but a deeply selfish part of me wanted Wolfgang to know all the things I felt, I wanted him safe with me at my remove. I made this child. He was mine. I wanted us to be angry together.

I call Andrew from my hotel in Palm Springs.

"Wolfgang claims," says Andrew, "that he misses the past. He wants to go back to before he knew about other people's minds."

Wolfgang had reached that developmental stage where children begin to truly grasp a Theory of Mind. He'd started to infer that other people's behavior was connected to mysterious, hidden thoughts, feelings, beliefs. All he saw now were minds. Every object had one, balloons, food, blades of grass—everywhere, thinking, feeling, unknowable.

I remembered myself at his age and the anxiety that had defined me then, the fear I'd felt driving around massive Midwestern parking lots. I was looking to my mother and to my father, looking to see what they hid in their private minds. Was I a burden?

Wolfgang looked to me to see what I hid in my private mind and what he saw was the assumption, always, that strange minds were cruel.

But what had been in my mother's mind? Happiness to have me

with her. And what did Wolfgang see when he looked to Andrew's example? Empathy.

When the tall tennis player had passed me and said, *What is that? What the fuck is that?* and when he'd mimicked my walk and when he'd laughed at me, I'd felt relief. It's not me ruining things for Wolfgang, it's the shit *out there*. I am right, I am right, I am freed of responsibility, because, see, didn't you hear that guy? I am right to be angry, I am right to be afraid. And that was how I lived my life. And maybe that was a good enough life for me, but it wasn't a good enough life for Wolfgang.

In the late afternoon, I decide to take a walk. I see a crowd by a practice court. I find a seat in the stands and I see him. He's right in front of me, Roger Federer, for the first time. The moment hangs suspended as if it caught itself on a nail. He's hitting casual groundstrokes with Lucas Pouille, a young Frenchman on the tour, but poor Pouille is irrelevant. Everyone wants Roger. All around me people look like they have phones for faces. I see a large red sign held aloft at the edges by two fans. It has a Swiss white cross on it and reads, "SHHH!! GENIUS at WORK."

After a while, I notice two women gesturing to me. They've seen my press pass. "Let us offer you an exclusive interview," says one.

Before I can ask a question, her friend hisses, "His game is so beautiful!" She leans in to share this news with only me. "So beautiful," and then says it a third time. She points out to him playing in front of us and searches for a further thought, her mouth open. She keeps pointing.

"He's the classic player," says the first woman, working to present herself to me as the more serious of the two. "He can serve his way out of a jam. He can just execute."

"He dances on the court!" says the friend. "He's a dancer. He glides.

When he plays it's just beautiful, so beautiful. To see his strokes up close like this. To see the smoothness of his strokes . . ." and her voice trails off as Federer's just hit an effortless crosscourt forehand and she disappears momentarily into her observation of him.

We watch him quietly until the serious one regains her composure and says to me in her serious voice, "A lot of people can play, but will miss when it matters. Being able to remember your mechanics under pressure: that's the key. People collapse under pressure. But a genius meets and solves the moment."

I observe Federer in a trance; my mind loosens in the heat of the midday sun. The intricacy of the human body comes into focus. So many minute actions have to perfectly align to hit a ball the way that he was able to hit a ball. His movements pure, efficient; his body, a collection of disparate parts working together with unified intention. When people looked at him, they saw a capacity, and when they looked at me, they saw a lack.

I become irritable. How absurd it is that we are all gathered here to worship a man who can hit a ball over a net. I feel foolish and angry. Why did he get to be the genius, exalted for something as meaningless as getting clean strings on a fast-moving ball?

But then the ball arrives. Federer sees it and lifts his body in the air, turning effortlessly; his arms unfurl like two waves moving in opposite directions and he hits his one-handed backhand. As he extends through the shot, his chest opens wide and his arms keep reaching, away, and the movement ripples down through his fingers, which are so relaxed that they look weightless, fluttering briefly in the breeze, and it is beautiful, so beautiful.

When I return my attention inward, I find my irritability is gone. The moment of beauty I'd just witnessed took me out of myself and now my self-centered thoughts are replaced by the electric pleasure of seeing someone be so good at something. Kant wrote that a genius could create something that displayed the furthest edge of our humanity and, by doing so, give us a sense of what might lie beyond that edge. Genius opened up a space for God to encounter man.

Roger Federer is, on this day, the best male tennis player in the world. He explores the boundaries of human ability right in front of me, tests the limits of perceptual processing, of movement, of harmonious interplay of mind and body. He opens a liminal space and ascends right to its edge, gifting all who watch a piece of crucial knowledge. I'm there to see it, and I do feel something like religious ecstasy, but not because I long to understand God, but because I long to better know what it means to be human.

Federer's feet flutter, then plant. He hits another backhand. A winner right past his opponent. The serious woman opens her mouth to comment, but I don't want to hear from her anymore and so I look at her excitable friend and smile.

And she says, to me, "Can you imagine doing anything, anything at all, as well?"

Federer plays a match and wins, then plays another and wins again. It solidifies something Federer clearly already knows, something he emanates in his every gesture. He's summited a new peak. He's aglow in press. He walks through a room as if he's never in his life feared the arriving moment.

I catch, from the corner of my eye, a head of white hair, and when I look over, I see a man who looks just like Geoff Dyer. He's sitting in the front row, watching Federer with a reverent attention I recognize as my own. I keep looking back at the man. He looks exactly like the actual writer Geoff Dyer. When the press conference is over, the man stands, and I follow him. I follow him down the hallway, through a door, down a second corridor that leads us back outside. The hot desert air envelops me and I feel sleepy. There's a sudden thinness to reality.

I'm certain that I'm dreaming. I've fallen asleep while reading Geoff Dyer's travel essays and surely I'm in bed, back at the hotel, maybe just waking, my semiconscious mind imagining him walking just ahead of me. Or maybe I'm dreaming this whole tournament. Of course, no magazine, let alone *GQ*, has simply allowed me, a nobody, to be credentialed in their name. Suddenly this whole experience reveals itself as a mirage, full of figures conjured up by a deluded mind, fantasies dancing before me, but remaining out of my reach. This white-haired man floats ahead, reenters the press office, moves through the glass doors, and takes a seat alone in the press box. He looks out at the empty court. I stand behind him and wait to wake up. If I tap his shoulder, it will be a foreign face, I'm sure, that turns to me. Or maybe it will be my father's face. Maybe it is him I'm dreaming of. I walk away from the door, toward my desk, but then I return and stand behind the man again and finally I hear my small voice say, "Excuse me," and the man turns to face me and I am awake and it is Geoff Dyer.

The first thing I say is, "I can't believe you're here." He tilts his head to the side and smiles. "I mean," I say, "I'm a big fan of your work."

He thanks me politely, not looking past me or around me to an exit. I ask him what he's working on and who he's writing for and he laughs and holds up the press credential hanging around his neck. It reads "Geoff Dyer—*Palm Springs Life Magazine*." He shrugs.

"I'm here with my friend from this magazine. I live in Los Angeles and I wanted to watch some tennis," he says. "Also, I just like to be in proximity to Federer."

"Me too," I say, and we talk for a while about the presence of Federer, how he walks, how people around him sit up a bit straighter. Dyer tells me he wished to have dinner with Federer and that he thought they'd be friends. "I think we'd get on well," he says. "He clearly has a sense of humor. It's part of his charm."

"I think he'd get on well with me, too," I say, and Dyer scoffs a bit, but not begrudgingly.

"Well, that's the thing about charm," he says. "It convinces other people to believe that they might be special, too."

"I saw you in his press conference," I say. "Will you ask him a question?"

"Oh, no," says Dyer, laughing, "I wouldn't dare."

I sit in the front row. I pretend no one is behind me in the conference room, no one is near, it is just me.

Roger Federer enters and takes his place behind the microphone.

I signal to the official that I have a question.

The official indicates that I'll speak third. The first person asks their question and I hear nothing but a rushing in my ears. I silently practice the wording of my own question in my mind. I do not feel confident, I do not feel calm, but I do feel able.

Roger Federer finishes his answer to the first question and now the second one is being asked. I feel light-headed, my chest is tight. I know I am making things worse than they are, but this is deep in the core of being a person. You are the thing standing in your way.

His answer to the second question ends and then the eyes of the official are on me and it's my turn and with the official's eyes, so follow Roger's eyes, and he's looking directly at me and only me, waiting for me to ask my question, and I realize that the question I've been practicing is gone, but more than that, my grasp of language altogether is gone and I wonder if it's possible to just walk out of this room, never to see any of these people ever again, but instead of leaving I open my mouth and I speak.

When I return to my desk, the French and Australian journalists are sitting at theirs and they turn and acknowledge me with a nod.

"Nice question," says the Australian.

The French journalist agrees. I want to ask them for a group hug. I say that I'd been very nervous.

"Sure," says the Australian, "but you'll get used to it. Soon this will all seem very normal to you."

"I've been scared of everything for days," I say. "Scared even of you two."

"Us? Come on," says the Australian, rolling his eyes. "Scared of the janitors, too? The food vendors? The lizards scurrying past?"

Later, I see Geoff Dyer in the press box again. He greets me like an old friend. I tell him about the press conference.

"Were you afraid?" he says.

"Yes," I nearly shout. "So afraid I lost my English."

"Oh?" he says, surprised. "What is your native language?"

"English," I say, and he laughs good-naturedly at me and I laugh at me, too, but we both know I've done something he could not do.

We all sit together in the press box and watch the championship match, which is won by Roger Federer.

"He made that look easy," says the Australian, whose name is Matt.

Does it matter, I ask Matt, that Federer makes his wins look effortless?

"Tennis is not easy, it's frustrating, it's physically taxing and you're constantly failing," says Matt. "And life is hard, too." We both look out at the crowd. The stage for the trophy ceremony is being set up. "So, isn't it just lovely when things just work? All these difficult bits of life—just flowing, easy? And we watch a player at his best and almost forget for a moment how hard everything is? We get to sit here in the sun and see something a little more perfect, a little more beautiful."

The Kantian man looked at *threatening rocks, thunderclouds . . . the boundless ocean heaved up* and found the courage to believe his reason could take on nature's power. This assertion of rationality and the individual will puts the Kantian man in opposition with all the Grand Assimilating Systems—Religion, Nature, Society. This is the basis of scientific progress and the analytic philosophical tradition.

Iris Murdoch, in a critique of Kant, writes that this "Kantian man-god" sees "the sovereign moral concept is freedom, or possibly courage in a sense which identifies it with freedom, will, power." By asserting his separate self, he puts himself on a path that, step by step, leads him upward, until he is above it all, utterly alone, alienated. This man, writes Murdoch: "free, independent, lonely, powerful, rational, responsible, brave, the hero of so many novels and books of moral philosophy."

The experience of the Sublime, Kant thought, was an experience of the exaltation of the self against omnipotent nature. But Beauty was something separate from the Sublime.

Beauty, Murdoch argues, gave us an opportunity for an "unselfing." She writes:

> *I am looking out my window in an anxious and resentful state of mind, oblivious of my surroundings, brooding perhaps on some damage done to my prestige. Then suddenly I observe a hovering kestrel. In a moment everything is altered. The brooding self with its hurt vanity has disappeared. There is nothing now but kestrel. And when I return to thinking of the other matter it seems less important.*

I search and find Federer one last time on the practice courts. I will not leave here a brand-new person, but a piece of an endless puzzle falls in place. I think of Murdoch's belief that beauty has "unselfing" powers. By paying attention to beauty, I could break free of myself, clear my mind of selfish care, change my consciousness, and tune it toward a life with others.

I take a seat in the first row. Federer tosses a ball into the air and his body floats up to meet it. The ball then appears to have been convinced, rather than struck, by his racquet to graze the upper corner of the service box and bounce beyond Pouille's reach. An ace. Next to me, a true Lady of Palm Springs, drunk on Moët, wearing a visor and a tiny black romper, shouts her love out to him. She extends her arms and legs, spilling her champagne down her front. One of her wedge heels comes loose. A man behind her, presumably her husband, hands her his full, plastic Moët champagne goblet and she sips.

"I love you, Roger," she shouts. "I love you so goddamn much." She nudges another lady near her and that lady lets out a long "Whoo" sound. "Come on!" they say to others around them and soon a whole section of Palm Springs Ladies are swooning and cooing sweet things out to Roger. He and Pouille continue to play an easy practice set during which a ball is shanked off the end of Pouille's racquet and goes into the stands. A young girl passing by the court catches it and is uncertain whether she's allowed to keep it or must throw it back.

"Keep it," yells the lady in the romper, above her in the stands. "Fucking keep it. Roger Federer touched that ball. That's Roger's ball. Keep it or give it here." The girl looks unsure. A coach on the court extends his hand out to the girl.

"Leave her alone," screams the lady in the romper. "Let her keep it." Then to the girl she snarls, "Don't you dare throw that ball back." The girl runs off, the ball tight in her hands.

The lady in the romper stands, blocking her husband's view, and she starts to dance. She gestures wildly to all around her and soon the chorus of women have all resumed their "I love yous."

And me, I've never been happier, it really might be true, than I am right now. I feel the awe of proximity—to Federer, yes, but more so to the crowd. I'm swept up in their good feeling, in their total embrace of outward-facing enjoyment. With the Ladies, I'm free, I'm a mere speck in the crowd, awash in communal appreciation of beauty. Me and the Ladies, we are together in this. We will wait and watch, together, until Roger leaves the court. We will shout and coo and applaud. Gone are the aches from shivering hours on end in the overly air-conditioned press office. My muscles warm and relax, release, my eyesight sharpens. I am woozy, euphoric. The women hoot and so do I. In between games, Federer changes his shirt, exposing his soft skin and the Ladies erupt in high-pitched pleasure. My voice joins in, but I can't hear it; its uniqueness is absorbed by all the other voices.

10

Miami Beach

My mother is hungry.

"Order whatever you want," I tell her.

"I'll just go upstairs and eat some nuts I packed," she says.

"Ask them for a menu," I say.

"Ask who for a menu?"

I point to some guys in sunglasses standing behind a tiki bar, the roof of which is thatched with fake, plastic grass. In a flash, I am back on the Mekong with Chetra, lying on my back in the huts that stretch out over the muddy river. I listen for the phantom sound of the village boys swimming over with bags of mangoes for sale. I close my eyes and keep them closed, imagining myself elsewhere, but then I catch myself. I sit up and open my eyes. Wolfgang is swimming.

I'd spent the rest of the spring and all of the summer writing about tennis for *GQ*. Matt, the Australian journalist, had been right; I'd gotten very used to being on the tour. Press conferences were no longer terrifying, and no one remembered anything I asked anyway. It all felt normal now, a job, which meant I'd stop doing it soon to learn how to write about something new.

It is autumn, three days before Wolfgang's seventh birthday, which we'll celebrate here, at this oceanfront hotel in Miami. We've managed to drag my mother along, too. It is our first true family vacation.

"I'll just go back to the room and eat some nuts I packed," my mother says again. She's happy to be with us but also a little miserable. She does not want to go on vacation to relax, she wants to go on vacation to ride more horses, preferably hard-to-ride horses on a treacherous trail in Montana or New Mexico.

We sit stiffly in pool chairs. I signal for a menu and a waiter walks one over but hesitates when he sees my scowling mother.

"Mom," I say, nudging her. She limply reaches for the menu. She flips it back and forth in her hand, reading both sides several times.

"I'll just go up for the nuts I brought in my suitcase," she says.

"OK," I say.

"Fine! Fine!" she says, and she orders a coconut smoothie. She watches Wolfgang. He flails about in the shallow end.

My mother's smoothie arrives on a silver platter—this and the pile of manicured fruit arranged on top horrifies her. She flicks the paper umbrella aside like a gnat. "What is this?"

"Your coconut smoothie," I say.

"A ridiculous coconut smoothie," she says, correcting me.

"You can't bear it."

"No," she says.

"Not a chore in sight."

"You're making fun of me," she says. She sips her smoothie. "You think I'm too country for this nice hotel."

"I think you can only enjoy things acquired through arduous labor. There is a wheel in your head that can't stop spinning."

"I finally stopped looking around for papers to grade. Took about ten years of retirement, but I finally stopped looking around for them. Now you're the one always grading papers."

It was true. I had a stack on my lap but had stopped working after I'd lost my pen to the crevices of my lounge chair.

Wolfgang floats by on a white unicorn. Another kid swims over to Wolfgang and grabs him by the leg and flips him off the floatie. A previous me seethes with rage, ready to hate any stranger in our midst.

But now I think only of Wolfgang, who is completely fine, but bobs up, eyes wild, searching for me.

"Help! Help me!"

"You're OK!" I say. "Just swim, try to swim!"

Wolfgang is too scared to see how safe he is, still in the shallow end. All he has to do is straighten his legs and he'll be on solid ground. But he's lost to panic and starts to sink. Andrew shakes himself free from a half nap, levitates above his lounger, dives in.

My mother says, "Not all songs need covers."

The resort has hidden speakers everywhere. Reggae "Karma Police" is playing.

She says, "I need a break from"—she gestures to the palm trees—"*the party.*" Reggae "Take on Me" plays next. She says, "This nice resort needs to learn to leave some songs alone."

"Mom?"

"What?"

"Where are all your chores?"

Her horse, Jimmy, had died the week before. She says, "Jimmy couldn't eat hay toward the end. I'd mix him up beet pulp. I'd cut a crack in a carrot and shove his pills in that way. I made his dinner special every night and took it down to him and tucked him in and now he's gone and I'll never get to do that again."

"I'm sorry about Jimmy, Mom."

She says, "I knew Jimmy wouldn't last forever. I knew that." She looks at me, lost in a thought. Her eyes close and her face falls into a deep frown. "It is awful," she says, shaking her head in disbelief, "what they've done to this song." The hidden speakers play a reggae rendition of "Walkin' After Midnight."

We towel off and return to our room to rest. I switch on the TV and flip through channels. Andrew and Wolfgang cuddle on one of our two hotel beds and drift off. My mother and I lay awake on the other bed.

"It's light out," my mother says. "Shouldn't we be doing something?"

"Wanna watch *CSI*?" I say to my mother.

She says, "I don't like murders. Shouldn't we take Wolfgang to the park?"

"He's asleep, Mom. Wanna watch *Law & Order*?"

"I don't care for murders."

We settle on a singing competition.

"My god," she says. "So much HAIR on this stage."

Endless commercials. In one, a boy comes home from college and his mom is so excited to see him, but he keeps being pulled away by one thing or another and then leaves his mom alone so that he can go hang out with his friends and the mom is sad. I look at Wolfgang. There's a halo of sleep sweat on his pillow. His mouth is wide open, his lips twitch as he breathes a heavy breath, in and out. He has one leg and one arm slung across the torso of his sleeping father. I make a strange noise.

"What are you doing?" my mom whispers.

"Nothing."

"What's the sound you're making?"

"Dramatic sighing, I guess."

"Stop that." She pokes me. "What's wrong?"

"I'm Wolfgang's favorite person right now," I say. "I don't want there to be a day where I'm not his favorite person."

"You mean you don't want him to grow up."

"He's going to go to college and he's going to think I'm so boring."

"Oh, god," my mom says, rolling her eyes at me.

"Were you sad when I went away to college?"

"Of course."

"You didn't say a thing when you took me to Boston. You just dropped me at my dorm and then were like, 'See ya!' "

She says, "I was *not* like, 'See ya!' "

"You left me in a big city so far away and you were so stoic about it."

"Well, I was very scared."

"You kept that from me. You hid that."

"Yes, because I didn't want *you* to be scared."

A taxi takes us to dinner. The driver just arrived from Haiti and had been sleeping in his car until a week ago. My mom asks him questions. When we get out, she hands him a wad of cash.

"Mom," I say, as we walk toward the restaurant, "I already paid."

"Everyone likes a little cash," she says.

A man passing us on the sidewalk hears this and asks my mom if she has change for a twenty and she says, "Let's . . . see," and opens her purse and separates out all her bills slowly.

"No," I say to the man. "No change for a twenty," and I shuffle my mother through the restaurant doors.

"You're so cynical," she says. "I didn't raise you to be so cynical."

When the waiter arrives at our table, my mother says, "Did you throw out your romaine lettuce? Romaine is giving people E. coli again."

Our waiter nods mournfully. "Poor romaine," he says.

"It's the second time!" says my mother. "It was giving people E. coli before and now there's no trust left. Romaine will never recover from this."

She and the waiter discuss their favorite lettuces for five full minutes and then, at the end of our meal, he brings my mother a free dessert.

On the way back to the hotel, a different driver tells us another sad story: her husband abandoned her and then her mother got sick and so she moved back to the worst city in the world, which was Miami in her opinion, and then her mother had died last Tuesday.

My mother listened and asked questions and, before they parted, my mother left the driver her boxed-up free dessert and a fistful of cash.

"Look at this!" my mom says, waking Wolfgang the next morning by placing a chocolate croissant on his bare chest. She's been up for hours, has discovered the best bakery in Miami, and has returned with treats for everyone.

Wolfgang lifts the croissant by one edge and looks at it skeptically.

He says, "Does it have peanuts in it?"

"No, no," my mother says.

"Peanuts will kill my dad."

"We know, we know," my mom and I say.

"I don't eat peanuts on principle," says Wolfgang.

"We know, we know," we say.

My son says, "Dad saved me from drowning and so I don't want him to die."

"You weren't drowning," my mom says. "You were just in a little over your head."

Wolfgang pouts.

"Are you going to feel sorry for yourself or are you going to eat a croissant?" asks my mom, sitting at Wolfgang's side.

He observes the pastry, turning it over in his palm. Then he takes a bite, chews, takes several more, and says, "I've never had this before, but now I know"—he wipes chocolate from his mouth—"it was the treat I've been missing all my life."

We eat lunch outside at a café. A lady walks up from the street.

"I'm harmless," says the lady.

"We have no cash," I say. "None at all."

"I haven't eaten for days," the lady says.

"Here," my mom says, "take our food!"

"No, I'm thirsty," the woman says. "So thirsty."

"Here," my mom says, "take our water!"

"Could I have five dollars instead?"

My mom pulls out her wallet and murmurs, "Let's see—"

I put my arm across my mom.

"We have no cash," I say. And the woman splits.

"Why not let me give that woman some money?"

I say, "You are out here telling the world to take take take."

My mom says, "That woman needed help."

"She's scamming you," I say. "Why do you always let people do this to you?"

"But you don't know," my mom says. "You don't know for sure that every person's motives are take take take! You can be so hard on people."

Andrew and Wolfgang eat their sandwiches and watch a video on Andrew's phone.

My mom says, "One time, I was in a hurry and I didn't have any cash on me, not one cent, and I was hustling around outside the hospital waiting for your granddad to get some tests done, and I was running to the car and I found one quarter on the floor and I used it to feed the meter so I wouldn't get a ticket, and then I was running back to the hospital and I passed a guy who was sitting on the sidewalk, hand out, asking, and I didn't look him in the eye because I was worried about Granddad and was racing back to the hospital and the guy yelled at me and he said, 'You don't even acknowledge that I'm here! I've got all these struggles and you treat me like I'm nothing, like I'm trash someone else will pick up,' and I yelled back, 'You don't know a thing about me. You don't know a thing about my life. I'm outside of a hospital, you don't wonder why!' and he said, 'I'm sorry, you're right, you might be in pain' and I said I was sorry, too, and we talked for a while and left as friends."

"He was nice just to get your money," Wolfgang says without looking up from the video he's watching. There is no way to speak without him hearing, no place to hide where he does not see me. I cringe. My mother looks at me and then looks at Wolfgang. She pulls him close and kisses his head.

"I didn't raise you to be so cynical," says my mother to me.

"But you did! Remember you told me to just play my card and get what I needed from people and move on?"

"That's one thing I said in a lifetime together. You really took that and ran with it. Get some balance. You have a big chip on your shoulder," my mother says to me.

"Of course," I say.

"You say it like you're proud of it. But you're only burdening yourself, carrying around all that anger and resentment. I wonder what it would take for you to lay it down."

I wake the next morning before dawn to the sound of air rushing in from an open window. My mother is dressed and sitting quietly in a chair in the corner of the room. She stares out into the early morning darkness, likely looking for a barn to clean. My son and husband snore away. I pull on clothes. She and I walk without speaking out of the room and down the hallway, down the elevator, and into the lobby where coffee waits in giant urns. We walk outside, all is quiet.

"Thank god, they turned down *the party*," my mom says.

Submerged pool lights turn the water's surface a ghoulish neon green. Some men crouch at the edges of the pool, sweep out its leaves. They wave to my mother and she greets them by name. We cross the courtyard and walk toward the ocean.

"How long is this boardwalk?" my mom asks.

"Very long," I say.

"What a nightmare." She's stopped and is looking at the wooden planks that make up the boardwalk. She counts the number of nails in each plank: eight.

"Some poor guy," she says.

"Who?"

"The maintenance guy who's gotta pull up all these nails to fix the boardwalk when the boards need changing."

"Who?"

"The someone who fixes things," she says. "The who that *fixes* things. The person who works. The world is not made up of resort *guests*."

There are heavy clouds hanging over the water. The sun begins its ascent by first tracing a line across the horizon. The beach is empty. It is all ours. Waves roll in and out. We sit in the cold sand and drink our coffee. My mom counts all the shades of blue she sees.

"Dad wrote to me recently," I say.

"Really?"

"Well, a year ago, actually. It's been a year already."

"Is he OK?" my mother asks.

"He is in a seminary now."

"A what?"

"In a seminary."

"Doing what?"

"I don't know."

"What kind of job is he doing?"

"I don't know."

"He just said, 'Hi, I'm in a seminary now'?"

"He wrote that the only happy memory he had was driving around in a car with me when I was seven."

"I believe that," my mom says. "Driving around, talking to you, listening to your batty kiddo ideas, that was a true thing for him. He loved talking with you. He just couldn't figure how to—"

"How to what?"

"Stay in it. Stay in the moment with you. He was always sure something was missing and that he'd find whatever it was in the next thing and the next thing."

"Maybe he'll find it in the seminary."

"Right now, you are your dad's age when he left for the last time and Wolfgang is the same age you were. I think about that. Do you think enough about that? You and your dad have a lot in common."

"In what way most of all?"

"You both love to pull a quote. You like concepts. You have a knack for justifying anything you do by appealing to ideas."

"But you don't do that."

"I do. No one is exempt. The reason I didn't leave your father was that I was sure the right concept was for a family to stay together. I chose a concept over a more specific truth. But I thought maybe I couldn't . . ."

"Couldn't what?"

"Couldn't raise you on my own. So, I appealed to an idea—the ideal family—and it was a helpful concept to cling to, but it wasn't right."

"I think if I hadn't been born disabled, if I'd been born the child he'd wished for, then he would have stayed."

"That's absolutely false. Wildly false. This is your big problem. You're always looking at the wrong hurt."

"What do you mean?"

"You have to try to ease up on the brooding. Most of the things you're hurt about have nothing to do with you. Your dad leaving had nothing to do with you. He was the one with limitations, not you. He was afraid."

"And what about you?" I ask, finally.

"What?"

"Were you afraid? You're always saying that you weren't afraid when I was born, that you weren't worried about me, but I don't know how that's really possible. I have a hard time believing it."

"You think I'm not being honest about that?"

"Not on purpose, but, I mean, for women there's no bigger concept than that you're supposed to love your child. You're supposed to want your child. So, do you think you're sort of dulling the truth of your memories with that concept?"

"No, no. When the doctor handed you to me—"

"And I was all twisted up—"

"Don't interrupt with your own version of the story. If you want me to tell you, stop butting in with your own ideas."

"Sorry."

"Look," she says.

"I'm sorry, Mom."

"The doctor handed you to me and I saw this beautiful healthy child, my child."

"You did?"

"Yeah, you glowed, you were glowing, really."

"You weren't scared at all?"

"No. It wasn't a scary moment for me. I just looked in your eyes and your face was so . . . You were so alive. I saw a light, a consciousness, you, my child. I wasn't afraid because I wasn't inside myself, thinking about myself. I was just right there with you. That's the truest thing I can tell you."

My father had an idea for a children's book. He recited the idea to me many times when I was little. It was to be a story about beauty. It begins with a father saying good night to his daughter. The daughter is afraid to be alone and so she begs the father to stay and teach her something new. *What do you want me to teach you?* the father asks and the daughter says, *Teach me about beauty. What is the most beautiful thing in the world?* And just then a snake appears at the window. In the Western tradition snakes got a bad rap. They were seen as symbols of evil and temptation. But in the Eastern tradition, snakes could represent growth, rebirth. The snake comes in through the window and takes the daughter and says he will show her the world. And the snake takes them to the Taj Mahal and the daughter asks, *Is this the most beautiful thing in the world?* And the snake says no and then shows them the

Grand Canyon and the daughter asks, *Is this the most beautiful thing in the world?* And the snake says no and takes her to the top of Mount Everest and then to the bazaars in Nepal and then to the temples in Japan and the daughter says, *Which of these things is the most beautiful?* And the snake says, *None of these are the most beautiful. I will take you one last place.* And the snake flies the girl back over the fields of Kansas and down a familiar street and through the window of a familiar home and the snake says, *Open your eyes and see*, and there is the father reading this story to his daughter.

My father understood a good story is a circle that finds the hero back where they started, but with new knowledge. He wanted this to be true. He also knew a good story ended with the hero realizing the world was bigger than he thought, bigger than himself. But the hero can't realize this until he returns home, and my father didn't have a model in his life for the return, only the quest, so he couldn't write the story because he could never get to the ending.

"Sometimes," I say to my mom, "I feel like I don't know how to return home."

She shrugs and gestures behind herself at our hotel that held my sleeping family within its walls.

"What a waste," she says. My mother wasn't patient with people who couldn't get their act together.

"When you talk about that feeling of looking into my eyes and seeing me . . . I didn't have that feeling when Wolfgang was born. It was a scary moment for me. I looked at him and felt really afraid."

We're quiet for a while, listening to the waves.

Then she says, "When I looked in your eyes on the day you were born, I felt connected—to you, yes, but it was more than that. I was connected to Motherhood, this complex human experience, and suddenly the world got really big. An aperture opened in an instant. But as the world appears wider, I become smaller in comparison. This can be difficult for people like you and your father, who see themselves as the protagonist of some grand story. You want to feel yourself grow bigger and the world grow smaller under your intellect. But that's a trap. You're just a tiny piece of—" She waves her hands out toward the ocean. "No matter what you've been through, your story is one among many. Try focusing on that," she says. "Focus on the feeling of the world getting wider."

Later that day, we watch Andrew swim out with Wolfgang on his shoulders to a sandbar. He flips him out into the surf and picks him back up and swoops him around his shoulders and tosses him in the air and dives him back under the surf. My mom feeds our lunch to the seagulls. Later, little coincidences fall in our favor. Doors open right when we approach. Tables facing the ocean free for us. We feel special, we hold on to this luck as long as we can, but it isn't ours to keep.

My mom reads out loud from Wolfgang's favorite book. I'm grading essays, but my mother's voice makes it impossible to concentrate. The rain comes in, no beach today, the sky is black.

My mom looks over and says, "Wanna hear about Rondie?" and then she reads a story about a girl who had twenty-two beautiful teeth. Wolfgang giggles on the bed beside her, head on her shoulder. In a few hours we'll take her to the airport. We'll stay one more day before returning to Brooklyn. There are busy months ahead, responsibilities to bear, so we know we will not see her again for a long time.

Our hotel room is small. Two beds, one bathroom, balcony, stucco ceiling. The room wraps around us tight, and I feel too aware of all our little personal sounds colliding. From my husband comes the muted scratch of fabric on fabric as he, an inveterate fidgeter, rubs one socked foot across the top of the other over and over. He's on the bed beside me staring into the void of a video game. His fingers move, buttons clack; Wolfgang sneezes and laughs; my mother scratches her nose with the heel of her hand—how many times in my life have I seen her do this strange gesture? Thousands and thousands—and she reads out loud. I try to erect a mental barrier to all this human movement and noise, it's all so close, constant; it eats away at my edges. I give up on my grading and stare at an article on my phone, but the sound of my mother's voice pierces the surface of my private screen, scrambling all the words.

My mom says, "Wanna hear about Mrs. Gorf?" and then reads another story from Wolfgang's book about a mean teacher with a pointed tongue and pointed ears. She raises her voice to a screech, does a Mrs. Gorf voice: "I do not allow crying in the classroom!"

I give up on the article and turn to Andrew. I bat his phone from his hands and put my head on his chest. He puts his arms around me, kisses the top of my head, and says something into my hair. He says that I am a nice wife.

"Start over," I say to my mom. "Start that story over."

She flips back a few pages in the book and restarts the story of the sleeping girl who flopped onto her desk and rolled out the window. I remember this one. This had been my book before it was Wolfgang's. She'd read it to me when I was his age. I look at Wolfgang, snuggled deep into my mother's neck, and I remember.

When she finishes that story, she closes the book and says, "I'm done."

"More, Yoyo!" Wolfgang shrieks.

Yoyo is the name he called her when he was learning to speak and could not say the word *lola*, meaning *grandmother*. Yoyo stuck. It is what we all call her now.

"Mom, tell Yoyo to read one more," he says.

He was born and changed my name and her name. He kisses Yoyo's cheek. Our balcony door is open, the rain is loud and the ocean rolls in and out, both perfume the air, brine and iron. My mother reads another story.

"More," we say when she finishes. I turn my face into my pillow.

"No, no," says my mother. "I'm tired."

"More!" says Wolfgang. "We are your children! Read to us!"

I press harder into my pillow, I can't bear to look at her, but I know she's looking over at me, I can hear her voice reaching out to me.

"OK," she says, "last one." I hear her sip something. It's a (ridiculous) coconut smoothie. She's come to love them. She takes a deep breath, opens the book, and the voice I know best begins to read and I listen to that voice and it braids through the sound of the rain and the waves, and then she's come to the end of another story, takes a breath, and when we're certain she's done, she keeps going.

Wolfgang wakes me before sunrise with an elbow to my eye. He'd left his bed sometime in the night and has crawled in close to me. His legs flip over my hip, his arms flail into my face, his breath is hot on my cheek.

"Wolfgang," I hiss. I am angry to be awake so early. I peel his sticky limbs from me.

From deep within the fog of sleep he mutters, "I love you."

"Wake up," I say.

He dresses. He walks sleepily at my side out of the room, down the hallway, down the elevator, and across the lobby, across the courtyard. Out on the boardwalk, I bend down and say to Wolfgang, "Can you imagine being the person who has to fix—" but he's long gone, sprinting off to greet a fit of blue, curling surf. Seagulls dive-bomb. Waves meet waves, whiten the sand, crash. The sun is inconsistent, rising differently each morning. When I'd come with my mother, the sun arrived like an angry thumbprint on the horizon, punching the clouds red. Today it stretches up from behind a gauzy curtain of ribbon clouds, easing itself into the sky with a whisper of pink. I miss my mother. I wish she was beside me again.

Wolfgang runs ahead of me on the beach, my boy, my unlatched shadow. I think of Andrew sleeping soundly seven stories up. He loved me knowing that I would not always be good or brave or happy or sure. He'd set me free on my own path, certain it would lead me to the right place. His faith in me, his selfless love—he gave it easily. And my mother, the queen of stark facts, had given her love easily and completely. She'd looked in my eyes the day I was born and she saw me.

My father wrote of the happiness he felt when we took our drives in the country—just us in our Ford 350, windows down, dust rising off the gravel road to sting our eyes, he and I, talking, laughing, happy, and then he'd written, "But Chloé was not enough for me. I abandoned her nonetheless." I wasn't angry to read this. I understood it now. I'd looked at a sonogram and seen a creature growing in me and thought, *Whose child is this?*

I was my father's daughter, and his future could have been my future had I not also been my mother's daughter and Andrew's wife and Wolfgang's mother. They built in me, bit by bit, a new capacity. I can be happy here, now.

The clouds above me show a shape, then shift to nothingness. Form gives way to formlessness. Wolfgang builds a sandcastle that gets pummeled by a wave. The sun rises higher. Around us, the inharmonious sounds of a new day. A beach raker rumbles awake and scrapes past; I hear the discordant chatter of more people arriving to see the dawn; next comes the *thwip thwip* of Wolfgang's feet flopping on the sand, kicking it up as he runs back to me; and in the sand: gnats, seashells, cigarettes, wiggling critters, carcasses, wings, feathers, seaweed, trash. Wolfgang falls to his knees, digs, finds treasures—shells and stones. Behind us, music, awful music. Our hotel has restarted the party.

At the end of my father's letter he wrote, *I quit drinking three-and-a-half years ago. I'm a slow learner.* He was changing. He could change and so can I. I feel something like forgiveness but wider.

Wolfgang and I walk to the edge of the ocean. The water rushes over me to my ankles, eroding the sand underfoot and I feel like I'm falling, but it's only ground below me shifting with the tide. We look out over the water. I'd wished for beauty to be a single, pure feeling ringing through me clearly, undeniably, creating truth, shining a beam so strong that it illuminated the entirety of my life. But what had come instead was a dense and drifting pile that carried with it a challenge: Could I see the salient thing? Wolfgang's hair, caked with sand; his bony shoulders shivering; his bright red gums, slick and swollen, new teeth straining, breaking through; his eyes shining, his gray eyes; his hand in mine. We'd not been given perfection, not godliness, not symmetry, nor gracious measurement, not a bad hand, nor a curse; we'd not been given anything other than a life to spend together; our lives, not easy or free from pain; we'd been given only a real life, dreadfully normal and sublime, and I would no longer betray its beauty by wishing it otherwise.

11

The Torqued Ellipses

Wolfgang stands beside me in the center of the room. Small paintings cover the walls. Above us, the museum's sawtooth roof lets light in through glass-paneled apertures. The way the roof rises in ridges makes it look like a claw snagged on the blue sky. This building was built to print Nabisco boxes and the roof was designed like this to shield the factory workers from the heat of the sun, but not its light.

We are back in New York, two hours north of our home in Brooklyn. Wolfgang is seven years and one week old. We are on a little adventure together, just the two of us, upstate for the weekend.

Wolfgang's hand in mine is hot. From where we stand, the art looks crude, small, unknowable; all fineness lost in the sprawling space. I approach a painting thinking only that it will be hard to look at more of them. It will take too much effort. The one in front of me is intricate, full of contradiction, or so I think at first, but I look at it and wait, I will myself to do the work of really looking until I have a thought about it, and then I wait until I have a second thought and then a third. I consider it with new patience, processing bit by bit the messages hidden in color and line and, after a while, an order emerges.

Wolfgang breaks free from me, unleashing himself with a jerk, leaving my discarded hand to hang in space, briefly, before it bangs against my leg. He sprints and slides on his knees, gleefully shrieking. His voice echoes off the walls. Then he's gone.

I hiss his name. I think I hear the thump of his sneakers in some other room beyond the door at the end of this endless gallery. I walk and walk; I cross the room at the highest speed my body allows.

My right hip hurts, wants rest, my right foot drags, I fall, I get back up, my wrist throbs, my hip throbs.

The museum is truly huge. One massive room opens into another. Vastness and vacancy disorient me. The only light comes in through the factory windows in the ceiling. Absent is the invisible hand of designed lighting, guiding, projecting a stabilized glow on an object as if to hold it in a halo that reminds us that what matters is mounted, lit, and labeled. Absent, too, is the rush of relief that comes under the fluorescent glare of the gift shop, signaling the end of the obligation to focus, and the permission to return to looser thoughts.

I make it into the next room just as a gray-haired guard leads Wolfgang back to me. The guard is stern, certain I am under his authority, and he says, "Miss, he can't just be running . . ." But then he stops, he's watching me walk.

Wolfgang whines. He's bored. He goes to a bench in the corner of the room, sits with his back to me, hunches his shoulders to affect a pouting posture, and I look at another painting, glad to be free of him. Then the shame that binds itself to that gladness arrives. Instinctively, I look to see if anyone is near, judging me.

I join Wolfgang for a rest on the bench. I reach for his hand, but he whips it away. He is old enough to know that all he needs to do to hurt me is withhold his love.

"Wolfie . . ." I say, but he shakes his head and then he stands up and runs, gone again.

I walk from gallery to gallery, seeking but not finding him. I'm annoyed first, then worried. The ache in my hip sends a sliver of pain shooting all along my right side. Where is Wolfgang?

"Did you see a little boy here?" I ask the gray-haired guard. He's kept an eye on us. He nods and points to a staircase. I follow it down, into a new room, which had once been the factory's freight car shed, but now as I stand at its threshold, it seems impossible this space has ever been anything other than a container—a membrane made of wood, concrete, and glass—for the four enormous sculptures at its center.

The other galleries had been open, too open, exerting their expansiveness on me all at once, but this one narrows itself around the sculptures, refuses a wide view, reveals parts when my eye desires the whole. A west-facing wall of windowpanes lets the light in at a slant; the floor is smooth and gray as river stone.

The sculptures are huge and moody; they loom over me as I approach them. Their actual shape is impossible to grasp in a glance. They are tilting cylinders of some sort, so large they nearly reach the ceiling. I circle the sculptures looking for Wolfgang. I call out his name. I follow what I believe are his footsteps echoing ahead, but I never arrive wherever he is walking.

In an interview, the artist Richard Serra had said that when he was designing these sculptures he'd been thinking of Japanese promenade gardens, which are designed to have hidden elements, revealed only by walking through them. What one can see of the garden depends on where one stands, and that view changes with every step forward. The garden cannot be taken in all at once; its reality is tied to a single viewer's specific position, but, as the viewer walks, this reality shifts, is proven to be necessarily fluid, changeable by space, time, motion.

Serra's *Torqued Ellipses* series, housed here in this former freight car shed, similarly reject a fixed view and, therefore, resist an immediate interpretation. He studied the Japanese concept *ma*, the idea that the space, the gap, the absences, between two objects, between two ideas, between sentences, words, breaths, held as much importance as the things themselves. The Serra sculptures contain a hidden second shape within them and viewers can walk a narrow path between the two unusual shapes. So the sculpture is not the hulking, twisted metal walls one could see and touch, it is the space between them. The sculpture is the absence, and the visible material, said Serra, is merely the "skin around the void."

I touch the Serras and they are rough, rusting. Their color changes as I move around them, first red, then amber, ochre, orange in the setting sun. Because the only lighting comes in through the window, everything I see is marked by time—by this solitary minute, by this hour, season, climate, by the cloud that crossed the sun just then. The patterns the light made on the sculptures would never be replicated, not exactly, and so when I look, I see the sculptures for the first and last time, the next time will be different, there's no way to keep hold of this moment, no way to revisit it, it exists and expires.

I call out Wolfgang's name and hear a scuffle; I call for him again, I'm certain he's near, but I can't see him, I move around one of the sculptures until its seam is revealed. This opening is a little mysterious, like a secret door. I enter the sculpture and see another tall wall within it. There's a narrow, mazelike path between these two walls, and I assume what I'm doing is walking between two concentric circles and so my mind projects expectations as to what this will look like as I continue to walk. I see my future set, but within a second or two it is reset.

The walls look unstable, twisting in odd misalignment. They slope toward each other and then, as I keep walking, they slope away. The distance between them changes with every step. I hear the thumping of little feet ahead and I think, *Ah, Wolfgang is playing in here,* likely running this narrow path like a racetrack. I stop and listen; he's ahead of me, I'm certain, but when I walk forward, something strange happens, the walls curve toward me, almost closing above my head, and the space where I stand is suddenly both so small and dark and Wolfgang is not there, he's not anywhere. The effect is disorienting, and everything gets a little fuzzy and dreamlike.

Serra's intention was to torque both the interior and exterior walls of the sculpture so that they were constantly shifting between leaning toward and away from each other. The shapes he designed turned out to be mathematically impossible; their angle of torque was too great, and every time he tried to have one of these walls made, the sheet metal kept splitting. Serra couldn't get the proportions right, so he started again. He was determined to make something completely new, to make shapes no one had ever seen, so that as one walked between them, seeing them in part, their mind could not correctly autofill the rest of the shape; any assumptions would, with each step, constantly be proven wrong.

Eventually, Serra found a shipbuilder in Baltimore who could make possible these impossible shapes. As I walk between them, my imagined path through—established in an instant by past knowledge and projected future—is continually thwarted. The light keeps changing. There is no stable way of seeing—what I can see is in flux, degrading, reshaping the exact moment in which I happen to stand there and as I keep walking and as time keeps going by, what I see keeps shifting, becoming duller or brighter or bluer in new light. I become more alert, more attentive. I stand closer to the sculpture, observing with increased care, trying to discern exactly what might change and when.

Inside the sculpture, there is only the present, unfolding unpredictably, and so the only pleasure comes from not knowing what is ahead of me. It is a child's pleasure, untainted, unafraid, sweet. I'd felt it before when I, as a kid, would walk in the woods behind my mother's Kansas farmhouse until I got lost. The trees I wandered between were thick and many. Their density disrupted my grasp of place and distance. I was so easily turned around in the woods, confused, thinking I'd walked a great distance when, really, I was walking in circles. The trees, their leaves and branches, broke and bent the light, deceiving me, making me believe that it was suddenly evening. When I'd come into a clearing, my path brightened, and it felt like the sun re-rising, day re-dawning. It went on like this, time changing at will, moving back and forth, until I left the woods and returned home.

From within the walls of the Richard Serra, I think of my own lost child. I whisper his name, *Wolfgang*.

I took the pregnancy test without fear. I knew it would be negative. I knew I couldn't get pregnant. Andrew and I set a timer and danced in the kitchen. The next day, I saw my son's face on a screen. A sonogram showed me his form within my own. My body had expanded to make a shape, an impossible oval, and within that shape was another body, breathing, tucked in a ball, my son, there he was, this second mind, second eyes and nose and mouth, and a second spine, but his was straight, not like mine. My life contained another. No one spoke. The doctor held the wand to my belly and allowed the moment to go on in silence, a kindness. I stared at the screen and felt myself passing between terror and awe and felt able to look up in observation of those feelings and, as the seconds passed, I felt the two walls of feeling bend toward me and then away, then toward me again. I moved between these two feelings but was not able to see beyond them. I watched my son on the screen and felt I was falling into the image, the center of which was his face, as foreign to me as the surface of the moon, and there I was, pulsing around him, my blood, my skin, wrapped around a void.

Six months before I became pregnant, I was in New York City stumbling out of a bar in Manhattan and into the arms of a friend. The city was new to me then; I'd only just moved there from Kansas, and I saw in it only a reflection of my own energy and desire. The city would unloose itself on me in the years to come, but then, in that moment, I loved it purely and without understanding. It was summer. I wouldn't even meet Andrew until September. My friend raised his hand and a taxi appeared and suddenly we were off, sailing across the Williamsburg Bridge, the moon pouring through the open windows, illuminating my friend who was, right then, the most beautiful friend in the world. I let my hand trail out the window and thought what a pleasure it was to rip through a summer night in a fast cab, the wind on my palm, on my face, in my hair, rippling it up off my neck. I closed my eyes and the moment slowed and I felt certain I had left the taxi and was weightless, above, swimming through the thick summer air. I opened my eyes and the world sped up, the skyline blurred, the headlights of other cars blazed past, above us only dizzy stars. My body was thrown forward into the back of the passenger seat and the taste of blood filled my mouth.

The taxi stopped somewhere in Brooklyn. My friend led me to a locked gate. Beyond the gate was a black void, an empty lot between buildings. A key materialized and we entered.

"Sit here and wait," he said, moving me to a bench in the middle of the dark lot.

He disappeared back through the gate. He was gone a long time, but I wasn't afraid. I was within the current of the moment. I closed my eyes and sank down into myself and realized that I felt no pain at all. I imagined my body splitting apart and my molecules spilling out to fill the empty lot. My friend reappeared with a brown bottle and we drank from it until I began to speak with boozy sincerity about all my dreams, my hopes, my plans. I saw myself owning little, needing little, staying free to make a life on insight, art.

He nodded. He put a hand on my back as if to say, *Really? All that? Will you do all that?* He was a dense shape in dim light. As we spoke, the sun slowly began to rise, and other shapes emerged. I began to see, slowly, that we were not sitting in an empty lot, but in a garden. He explained to me that the garden belonged to the neighborhood, that everyone could claim a space here to grow whatever they chose. What had looked in the dark like burial plots were raised garden beds. Around the sides of our bench were trellises, woven with vines.

The light grew and the scene sharpened. The more I saw, the more the air changed. It was suddenly filled with the scent of soil, and I heard the sounds of bugs and birdsong.

I felt I'd spoken the garden into existence—the more I spoke, the more the garden cut itself out from the darkness. And I believed that the future I wanted would arrive just like this. I'd simply plant the seeds and then the sun would rise on my efforts, bringing forth the bloom. I wanted to travel to every country and speak every language. I was certain I had art in me, books to write. I would be an intellectual, a poet, an adventurer, a T. E. Lawrence. There was no place in the arc of my imagined life for a child.

It was impossible.

I'd been told, over and over.

My body was incompatible for growing a life.

So, whose child was this?

In the days and months after his birth, I'd lie awake waiting for my son to cry. If he was quiet for too long, I'd feel a flood of certainty that he'd died. I'd get up in the gray light and I'd walk slowly toward his crib. I'd feel the weight of every footstep, each inch forward, knowing that when I crossed the room and peered into his crib, I would learn a truth I couldn't unlearn. I'd put my hand on his little chest and I'd feel it rise and fall and rise and I was relieved he was alive and I was resentful that this meant I had to take care of him and I was furious with myself for literally birthing into existence a love so strong it could wreck my life.

Some nights, I woke disoriented, unsure of whether it was dawn or dusk, and I waited, listening for the sound of him, and when it didn't come, I rose and walked to his room and put my hand on the doorknob, and I was sure my delusion would end and that when I opened the door I'd find not his nursery but my old office, with all my books and papers untouched, the walls unpainted, and then I'd rub my eyes, trying to erase the last images of a strange dream in which I'd become a mother.

My path within the sculpture narrows. These two ellipses, torqued at different degrees, making a shape within a shape, a distortion within a distortion. The steel sheet, a warm rust, the light, the redness of my womb, Wolfgang's red face, his body wet and raging in my arms, his voice calling out for me.

I think I am reaching a dead end and perhaps that is where Wolfgang waits, inside at the furthest point, for me to find him, and I step forward, certain I will see him now, or in the next step, in the next step. The space grows tighter, the walls turn in on me and the space becomes so tight that I can barely move, but I keep walking and the walls curve back away from each other, light streams in, and suddenly the path is wider. It is not a dead end as I'd thought, but a continuation, the path changed. But still, Wolfgang is not there.

For a moment I wonder if I'm dreaming.

Perhaps, I am about to wake up.

The more I consider this possibility, the more likely it seems that at any moment, I'll feel the hand of my friend shaking me, and I'll be back in that Brooklyn garden, back in a body incompatible for growing a life, the sun just coming up. Wolfgang will be a detail from a fading dream. Is that what I wanted? Is that my wish? Had it been my father's? No. I think he'd choose, if he could, to return to the dirt roads in the Ford truck with me next to him. And me, my wish is that I could somehow follow this path right back to my kitchen where my terrified younger self sat alone holding a positive pregnancy test. I wish I could reach my hand out to my past self, looking at that plus sign, and that I could say to her, *Well. What are you going to do now?* It was not a divine or magical voice, not an epiphany, it was my own voice, calling on me to face some hard facts.

At the center of the sculpture is a wide-open circle, Serra's void. The center is empty.

Wolfgang's laughter echoes around me. I know he is near.

Aristophanes in the *Symposium*: "Love is merely the name for desire and the pursuit of the whole." I had listened to this lesson too closely.

If love is the name of the pursuit of the whole, what is the name given to finding it?

I close my eyes and focus on the feeling of the world getting a little wider.

When I open my eyes, Wolfgang is there, laughing mischievously, running to me.

12

A Bar in Brooklyn

Days obey a new rhythm. I'd missed my students over the summer and show up to teach with a renewed sense of excitement. I attend all the dreadful faculty meetings, absently enduring my dean's scolding, and I'm added to several committees, one that requires me to fill out new piles of accreditation paperwork, and I can't stand this part of my job but I do not miss any more meetings and I do all the work. It is, of course, a small price to pay to make sure my family has what it needs. At home, we are happy.

One night, Andrew shakes me awake.

"Where's Wolfgang?" he says.

I run my hand across a pile of nearby pillows.

"He's not here. He's not in his bed?"

I search his room, call out his name. Andrew and I move through the apartment.

"Is he in the bathroom?"

"No."

Finally, we find him. He's wide awake, hiding under a pile of pillows on the couch in the living room.

"What are you doing out of bed?" I ask.

"What's wrong?" Andrew asks.

"I can see the fears," says Wolfgang. "They are everywhere."

"Oh, the spiders again," I say.

"Don't say their name!" Wolfgang whines. He'd recently watched a movie where spiders climbed out of a bowl of cereal and up the spoon and onto the arm of the protagonist and now Wolfgang is terrified of spiders. He couldn't believe that such horror could emerge from the everyday act of eating cereal and now he wondered where else it lurked. He thought he saw the fears in fabric, in our curtains, under the stove. The fears lived in the folds of the blankets I tucked around him every night.

I lead Wolfgang back to bed but promise to stay with him. Andrew, giving up on sleep, goes to the kitchen to make coffee.

"I'm scared," says Wolfgang.

"You're fine," I say. "It's just your imagination."

"My imagination is very, very scary."

"Picture something pleasant," I say to my son.

"Like what?"

"Like a doughnut."

He closes his eyes, then cringes.

"The fears are crawling out of the doughnut."

I get into bed next to him and tuck him under my chin. I stare at the part in his hair. So many hours spent staring at this head, breathing in the scent of his scalp, wishing to pull a particle of him back into me.

"Sleep," I say.

"I can't," he says. "I have to keep watch on the ceiling."

I look up and see stick-on stars above us, glowing an artificial green. We'd put them there together, Wolfgang and I, arranging them carefully, copying constellations.

"The stars are holes," he says. "The fears are coming in through the holes and will drop on my face."

I kiss his hair and start to doze.

"Mom?" says Wolfgang. "Mom, are you awake? You can't fall asleep and leave me here alone."

Andrew is in the kitchen. I can hear the soft sound of his coffee cup returning to the surface of the counter.

"I'm awake."

"I'm scared," he says.

I remember Wolfgang's worst accident. When he was three years old, he fell and caught a stiff bit of metal sticking off a fence on the way down. It cut him above his left eye. It wasn't so bad, but head wounds bleed. By the time we got him home from the park, he was covered in blood. The cut looked worse than it was because of all the blood. Wolfgang had been scared and cried and said, over and over, *My head is wet, my head is wet, what is happening?* He'd blinked blood away. He touched the blood on his cheek and screamed. He looked bad, so bloody, but he was fine, he needed a stitch or two. I couldn't calm him down. Andrew couldn't calm him down. We couldn't clean the cut because Wolfgang was screaming about his head being wet. I got an idea. I said, *OK, close your eyes*, and he did and I walked him over to the closet and my husband said, *What are you doing?* and I said, *OK, Wolfgang, take a look*, and he opened his eyes and he was standing in front of our full-length mirror and I said, *This is what happened to you, this is what it looks like*, and he leaned in and touched his face, which was smeared pink with blood and snot and tears, and he looked at the blood dripping from his wound, and he touched his bloody shirt, and then he said, *This is so coooooool.*

"Don't close your eyes," I say. "Just stare at the star holes in the ceiling. Look right at them."

"But the fears will drop on my face."

"Well, let's see. Let's watch and see if they do."

"OK," he says, but his eyelids flutter. He can't keep them open. He's so tired.

"I'll watch the ceiling," I say, "for ten seconds, I'll count, and then you can take a turn watching the ceiling."

"OK," he says. "Just ten seconds. But I'll count." And he counted *one, two, three*, and was asleep.

The predawn sky is gray; later, lilac. Then blue. I blink and suddenly the sun is up and the room is bright. I'm alone in Wolfgang's bed. Then I hear his voice. He calls to me from inside his closet.

"Ready?" he says.

"For what?"

"The big reveal!"

I hear clothes swooshing around. Hangers clang.

"Here I come," and the closet doors swing open. He steps forward, dressed as Death, scythe and all. "Trick or treat," Wolfgang growls.

Teachers line the school's vestibule dressed as various animals, superheroes or Harry Potter characters. They greet the kids with bags of candy, which Wolfgang eyes with interest. Andrew had drawn a jack-o'-lantern onto a pillowcase and Wolfgang holds it open, ready to receive. A kid crosses the schoolyard.

"Who are you supposed to be?" he asks Wolfgang.

"The Grim Reaper," says Wolfgang.

"Is that from *Minecraft*?"

"No," says Wolfgang, smiling, "it's from when we all die."

On our walk home, Andrew and I pass the same bodega as always. We pass our pharmacy, our hardware store, our café, and our bar. We go upstairs to our apartment, have sex, watch TV, unload the dishwasher. Outside our window, a familiar coo and splat. Pigeons shit on our air-conditioning unit. Andrew makes up a song about the pigeons and sings it to our cat. There's dinner to discuss. Frozen gyozas again. Andrew knocks over a cup of cold coffee, then cleans it up. He leaves for work and I grade papers. I pace around our apartment. I fold a pair of Wolfgang's tiny socks, too small by two sizes, and I put them in a special box with his other keepsakes in his drawer. Our cat sleeps.

After school, Wolfgang watches videos on his iPad while I sit on hold with the pharmacy. Mottled, tinny music comes through my phone's speaker. It fills our apartment. Wolfgang smiles, nodding his head to the sound. We are united for the moment in the rhythm of this waiting, the notes of the mechanical song. Wolfgang wiggles his chin as if it—and not some stranger's fingers somewhere, sometime on a piano—were producing the arpeggios.

Our friend Lincoln comes to live with us. He sleeps on our couch and recounts the details of his recent breakup. He moves cautiously through our apartment, touching door and drawer handles tenderly, stepping lightly, as if the slightest bit of force might also cause this home to fall apart around him. He plays video games with Wolfgang and bakes us ziti for dinner. He's been my good friend for many years.

Bobby still lives with us. His girlfriend Jess, the artist, often sleeps over. She, too, had recently gone through a bad breakup and is careful with Bobby, often withdrawn and distant, not wanting them to get too close too fast. She asks me how she can keep herself from hurting him. She tickles and teases Wolfgang so mercilessly and he laughs until no sound comes out of his wide-open mouth.

My old friend Becca had been living in Florida working a good job, dating a decent guy, and caring for a cat she loved but then the cat died and the job and boyfriend grew tiresome and so she quit them and Florida and called me and I said, "Move to Brooklyn" and she did.

Some nights we all sit together in the living room and talk and eat, and I find I can get through entire evenings without thinking I should be anywhere else.

On Lincoln's birthday, we go to meet some friends at a bar. Andrew stays home with Wolfgang. I show up late and see everyone already at a big table in the back. They're glowing, clearly a few drinks in. Some play pool badly. Everyone in our party is a writer or editor or otherwise tied to the publishing world and they leap at the chance to buy us drinks, which in turn entitles them to an hour's audience for their complaints about whatever it is they are working on. Those are the rules.

We sit in a circle, about ten of us, and scrutinize the merits of this or that book or article and we try to one-up each other on gossip about various writers and Lincoln is laughing and I'm glad. Lincoln brings up an article of mine that had recently won some recognition, and everyone briefly lifts their glass my way and nods.

Everyone but Kyle. He does not seem to register this little toast in my honor.

Kyle says, "My teeth hurt."

Kyle is a friend of Lincoln's, an acquaintance of mine, someone I exclusively see at parties, bars, or skulking around the edges of literary events. He is a handsome man with an incredible head of hair, really great black curly hair. I think his hair, rightfully so, had gotten him pretty far in life. It is fantastic. But, also, he is smart and well-read, and I'd been told he was a good writer. He had once worked at a hip literary press but was fired for reasons unknown to me. He is sitting next me, still talking about his teeth.

"Have you seen a dentist?" I ask.

"No one can see a dentist," he says.

"No one can see a dentist?"

"No one at this bar has health insurance."

"I do," I said.

"How?" Kyle asks. He looks at me with real surprise.

"Through my teaching job."

"Well," says Kyle. "I'm trying to write."

The conversation moves on. People pick apart a recently announced literary prize longlist. My editor from *GQ*, Kevin, is there and we talk a little about books and a little about tennis, but mostly we talk excitedly about my next assignment, which covers a totally new topic and is more ambitious. Kevin buys my drinks, and he buys Lincoln's, too. Kyle disappears for a long time in the bathroom and comes back glassy-eyed and distant.

I watch Lincoln for signs of sadness, this being his first birthday in many years without his girlfriend, but he looks OK, happy even. He tells our group that he is living with me for now and that Wolfgang had asked permission to make his birthday cake. Wolfgang had drawn in crayon a detailed blueprint for this cake, which was not a cake at all, but a dense mishmash of doughs—cake, cookie, and brownie. Sprinkles on top. Lincoln delights everyone at the end of the anecdote by producing a plastic bag filled with Wolfgang's treats, which we had made to his exact specifications. Everyone tries them and agrees they are delicious.

Each year, there's one perfect autumn evening in Brooklyn when the air finally shifts from dense and humid to crisp and clean and tonight is that night. The bar is busy but not packed. My free drinks taste great. My friends are here. But also, I realize, I am here. A normal person having a drink with her friends.

When it's time for me to leave, I hug the circle and kiss cheeks. I tell Lincoln to stay, that I'll make up the couch for him. I'm looking forward to the walk alone on this first night of fall. I'm looking forward to coming home. I'm certain I chose the right time to leave—buzzed but not drunk, sated but not full—and I know that leaving right then will close a circle on the evening, gifting me a clean memory, but then, on my way out, I pass by the pool table where Kyle is hunched, stick in hand. He is scowling, losing.

He straightens to say goodbye and we hug and I turn to the door. Then I hear him say my name. He asks if I'll talk to him for a minute in private.

"Sure," I say.

He'd like to tell me, he says, about hell. He's in hell. He's in indescribable agony that he then goes on to describe.

Kyle's wife—an artist whom I knew to be talented, hardworking, and kind—was enjoying an upswell of professional success. Kyle tells me her success has changed her into someone he didn't know anymore. Then she divorced him.

"I'm sorry," I say and mean it. I believe he is in real pain. I can see it on his face.

Kyle continues.

"There's more," he says. "And it has to do with you."

"With me?" I say.

Earlier that year, his wife, now ex-wife, had been diagnosed with a chronic illness. She was often unwell, bedridden for days.

"She wanted to leave parties early," Kyle says. "She got so tired. She didn't want to go to as many concerts. Sometimes she needed to take a cab and I resented her spending our money that way."

He tells me there'd been some incident with his teeth—he'd cracked one maybe and another fell out, I think, although I'm not entirely sure, I cannot quite follow the full saga of this man's teeth. Anyway, he'd come home to his wife, now ex, with these awful teeth and she was in bed in pain and he looked at her and thought he deserved all the pitying attention that evening but that he wouldn't get enough of it, that it had been usurped forever by his wife's illness.

His teeth, he explains, still hurt.

"It became a lot to deal with," Kyle says. "I'd come home, and she's in bed. She was so tired all the time and suddenly I had to take care of her."

This ex-wife of his is one of the art world's great beauties. And yet, when she was tired, when she was sick, she wasn't beautiful enough for her husband anymore. She was transformed into someone he didn't know, someone who was a lot to deal with. This, along with who knows what else, had built a rift between them and now they were divorced.

Kyle looks at me and pauses, searching my face for acknowledgment. I already know what he's going to say, but I'll make him say it.

"So, I was wondering," he begins, "if you could help me."

"How's that?"

"Maybe you could explain how your husband copes with the burden of your body?"

I walk home in the dark. The air is sweet with autumn rot. A cool breeze pulls goose bumps up on my arms. I think about what Kyle said and I think about the night a year and a half ago at another bar in Brooklyn with Colin and Jay, but mainly I think about the next article I'm writing for Kevin, about his excitement for it and belief in my ability to write it. I'm almost home before I return in my mind to Kyle.

A year ago, I might have taken his insulting question and turned it inward, using it as a weapon against myself. But tonight, as I listened to Kyle, I'd searched myself for anger or the impulse to retreat. But I found only attention. I had no need for my neutral room. It was his pain, not mine, and I would not take it on. It would stay with him. His sadness, I could see when he'd spoken to me, was destabilizing. He'd begun to come apart right in front of me; his voice breaking like the pool balls break, with a crack, a violent parting; and he'd cried a little, and below the quivering neon, the neon in moonlight, I saw a fragmented self in pain, I saw regret, self-loathing, ignorance, but I also saw his genuine desire to be understood by me, to be seen clearly by me, to feel a proximity to someone. And I could not give him that. But I could give him my belief that he was more than what he'd asked me. I could forgive him this difficult moment, see it for what it was, an intricate thing full of contradictions.

I was not always given the benefit of a person's second or third thought, and I was not always seen as whole, and this was not going to change; there would always be another Kyle, another indifferent man, another set of strangers in another alleyway, another Colin. But I could choose to see Kyle as whole. And doing so cost me nothing, whereas holding on to all the anger, anxiety, fear, and disgust I used to feel in the presence of others had nearly cost me everything.

What Kyle asked me has nothing to do with me. I reminded him of that, wished him well, and I went home.

Andrew is asleep but rouses when I get into bed next to him.

"Have I been a burden?" I ask.

"No," he says, laughing.

"Not my body, but me?"

"No."

"But I kept leaving."

"Yeah."

"And was that hard?"

"Sure."

"Are you mad at me?"

"No."

"Why not?"

"I knew you were on your way toward the person you wanted to be, and you'd get there if I just let you get there."

"But what if I didn't get there. Weren't you worried I might not come home?"

"No, I can't control what you do."

"You could have tried to control what I did. People try."

"Not people who think about their actions for more than a minute."

"So, you just thought about it?"

"You can't make people feel or do anything, least of all you. I can't even make you hang up your towel after you take a shower. No matter how many times I ask you to just hang it up, it always ends up at the foot of the bed."

"But you must have wanted to ask me to—"

"It's gross. It doesn't dry properly and gets a sort of, I don't know, not quite moldy, but mossy smell."

"I know. I'm sorry."

"I like your quirks, but not that one."

"Sorry."

"The towel thing."

"Yeah."

"But I can't get you to redirect that behavior. It would mean a lot to me if you would, but I can't force that. You have to want it."

"OK, I want to hang up my towel."

"Great, we'll see."

"Every time from now on. I mean it."

"I know you do."

"Do you think I can change?"

"Yes," he says, and I know that he is right because I'd felt so differently listening to Kyle speak than I had listening to Colin speak. I wasn't the same person now. I was oriented in a new direction. Kyle's words didn't find their way to me, not really. It wasn't that I thought less of him, only that I no longer thought less of myself.

"We've both changed," says Andrew. "Remember who I was when we met? I was eating one large pepperoni pizza a day and smoking too many cigarettes and playing video games all night. Do you remember who you were? We've both changed, but the difference is that I didn't know who I wanted to become, and you did, that was clear, and so I knew I needed to stay out of your way and just let you become the version of yourself you wanted to be, which I knew was, at its core, good. I knew you wanted to be good."

"How did you know that, and I didn't?"

"I can see you," he says.

"So, you didn't care that I left?"

"I care if you are happy. I'm in love with your happiness. My focus is on how you feel not what you do," he says.

We stay in each other's arms for a long time. Our breathing synchs. A car alarm, a siren, people shout on the street.

"You could have made me come home," I say.

"No, I can only try to be the person you want to come home to."

Wolfgang and I stand below a sulking sky. He's done with school for the day but wants to stay on the playground with his friends for a while before I haul him off to the bodega and then home. I let him have ten more minutes and watch him hurl his body around a jungle gym. He plays easily and happily with other children. Clouds form above. They make shapes upon which I can force a category—pig, shoe, hat—but the categories don't stick. Shapes dissolve; reality remains in restless motion; it was beautiful, it was Beauty, this engagement, this grasp and release, this brief but real discerning of the fleeting and how it redirects my focus from myself and toward the world.

We get our groceries. Wolfgang clutches a frozen pizza to his chest on our walk home.

"What shapes do you see in the clouds?" I ask.

"Just a buncha Pikachus," says Wolfgang.

After dinner, the three of us go out for a walk. Everyone knows us. The ladies at the pharmacy give Wolfgang a sucker. We get a bottle of water from the bodega and Wolfgang sits on the ground, petting the familiar cats. Nail salon, Chinese takeout counter, laundromat. At the edge of Prospect Park, we see a sign for a magic show.

"We should go," Andrew says.

"I don't know," says Wolfgang.

"What? You aren't curious?"

"Curious about what?" Wolfgang shrugs. "Someone doing tricks and pretending it is magic?"

"But maybe this time it will be magic," says Andrew. "Maybe it will be the best thing we've ever seen."

Wolfgang reluctantly agrees and we walk to a field where many young children, close to his age, sit in the grass in a circle around the Great Moody Trudy. I recognize a few of Wolfgang's friends from school and I tell him to go sit with these kids, but he refuses, and he stands in the back, at the edge of the crowd, next to me.

The Great Moody Trudy waves her hands over a fishbowl, drapes a cloth over it, says a few magic words, removes the cloth and reveals a fish swimming there. The kids clap and cheer, but Wolfgang is unimpressed.

"The fish was in the base of the bowl the whole time," he says.

The Great Moody Trudy starts another trick involving a newspaper. She pours clear water onto it, but when she rings it out, the water is green.

"The newspaper ink is food coloring," says Wolfgang.

The Great Moody Trudy balls up the newspaper, then launches into an impassioned speech about the environment and the importance of not littering. She leads her crowd, with a hand cupped behind her ear.

"Let's hear it, kids," she says. "This newspaper should not be left in the grass of this nice park! I should put it in the . . . ?"

And the children all shout, "Trash!"

And Wolfgang covers his face in disbelief and whispers, "Recycling! These idiots."

The Great Moody Trudy pours lemonade into a cup, starts to sip, then pretends to trip and out comes glitter everywhere.

"There was never lemonade in that cup," says Wolfgang.

She does a trick while juggling balls, one disappears.

"I saw her palm it," says Wolfgang.

"Tell her, Wolfgang!" I say. "Tell her you are not fooled." Andrew glares at me.

"I know how you're doing this trick," says Wolfgang loud enough for everyone to hear.

I pull Wolfgang close to me. He reaches up to hold my hand. We are united, safe, apart, against the scene, but together.

The final trick involves a stuffed rabbit going on a journey. The rabbit, we're told, goes into the forest of happiness and emerges with the key to a secret box that will unlock the real rabbit.

The Great Moody Trudy lifts a box and puts it on a pedestal. She pulls a drawer open to show it was empty.

"The drawer has a false back," says Wolfgang.

"Where is he? Where is the real rabbit? Is he in this drawer?" asks the Great Moody Trudy.

"No!" all the kids shout back.

"Yes, clearly, he is," says Wolfgang.

"Tell her she's busted," I say, nudging Wolfgang and we both crack up. "Tell her you got her all figured out."

Andrew grimaces. "Stop that," he says to me.

"Promise me," says the Great Moody Trudy. "Promise fingers in the air, everyone. Promise me you won't tell anyone if I reveal to you the secrets of my trick."

"I don't owe you that," says Wolfgang and we high-five.

"Chloé," Andrew says. "Don't."

"He's smart. Look how smart he is! He's figured out all the tricks," I say to Andrew.

"OK," says the Great Moody Trudy, "time to meet the rabbit," and she knocks on the false backing of the drawer and Wolfgang smirks and begins to say, "See, I told you . . ." but then the backing collapses and the whole box collapses and there is no rabbit there at all and Wolfgang stops and stares.

The Great Moody Trudy holds up the stuffed rabbit and says, "I guess there was no real rabbit after all."

Wolfgang's face falls. The Great Moody Trudy starts to put the stuffed rabbit back in her jacket and then with a flourish, she takes off her jacket, swings it around in the air like a cape, swishing it between us and the stuffed rabbit, and when she pulls it back, a real one sits sweetly in her hands. The crowd erupts, Wolfgang too. He's enthralled. He's seen something magical.

"How did she . . . I can't . . . Where was it?" he says, beaming.

"See! Not so smart now, you cynical little jerks," says Andrew.

The Great Moody Trudy waves the children forward, offering to let them pet the rabbit. To my surprise, Wolfgang darts forward to join the line and, without a trace of self-consciousness, stares up with a face full of wonder at the Great Moody Trudy and he tells her she is the greatest magician he's ever seen. He pets the rabbit gently, then nuzzles it with his nose, and the Great Moody Trudy leans down, and they embrace.

We stand at the edge of the park and watch Wolfgang play with the other kids.

Andrew asks me, "When Wolfgang was standing in the back of the crowd with us at the magic show, do you know what I thought?"

"What?"

"I thought," Andrew continues, "I went to school dances, and I didn't dance."

"Oh," I say. "Neither did I."

As kids, we'd both been compelled to derail experiences we felt excluded from. We'd believed and placed all our self-worth on the notion that to assert yourself against the crowd was a form of higher-level thinking, which sometimes it was and sometimes it was cowardice. We'd not been great at seeing the difference, but Andrew is better about this now and I am learning.

"I didn't participate," he says. "And it might have felt good to be united with my peers. But I missed the chance to find out."

"Right," I say.

"I look at Wolfgang and I see someone smart and kind," says Andrew. "I want him to be critical, but not at the expense of camaraderie. His compassion and his intelligence are complex enough to allow for both if we just don't—"

"Don't what?"

"If we don't get in his way."

Iris Murdoch wrote, "By opening our eyes we do not necessarily see what confronts us."

We are, she believed, anxious, self-absorbed, and we see through a falsifying veil of our own making that partially conceals from us the world.

We find places to escape from reality. *Where do you go?* my father had asked.

Our consciousness was often clouded in escapist tendencies, fantasies, reveries, and these retreats from reality are, she wrote, "profoundly connected with our energies and our ability to choose and act."

So, to make better choices and to be more just, we needed to change our consciousness and Murdoch thought that beauty was an agent of that change, but not through some appeal to order, proportion, ideal form, perfection. Instead, beauty helped us be attentive to a world outside ourselves. Beauty could help us improve the quality of our consciousness by briefly helping us "unself," gifting us a break from our "fat relentless ego." Taking in the beauty of the hovering kestrel gifts her a break from herself.

"The brooding self with its hurt vanity has disappeared. There is nothing now but the kestrel. And when I return to thinking of the other matter it seems less important."

One day bleeds into another. In the evening we go outside. It is the warmest it has been in weeks. The low sun casts a golden hue over everything. Halloween has passed, but people are still out in their costumes. A Dr. Strange vapes on the corner. Wolfgang's hair is slicked down with sweat at his temples. I hold his tiny hand and we walk down the street, two blown fragments passing through the ephemera of other lives.

A breeze, a glint of light. My son smiles up at me. Right then, at that second, he is content and safe, just on a walk with his mother in autumn, a walk he won't remember. But for me, there is nothing now but this walk, which will last a few more blocks, a few more minutes, the most boring, normal, and beautiful minutes of my life, and I think if I concentrate hard enough, I can get the minutes to leave their shadow on me. I want to stop the minutes from fading, I want to fossilize them and carry them off to be analyzed in my neutral room, but instead I pay attention and then I let the moment go and am rewarded, the next day, with new beauty: a morning song, a simple tune, the spatial rhythmic shuffle of Andrew in the kitchen in socks, the faucet singing, the tinkling melody of water running over the pots and dishes before striking the sink's metal basin, then a rinsing whisper, soap sloshing in the dirty coffeepot.

Acknowledgments

Thank you to my agent, Claudia Ballard, who changed my life for the better.

Thank you to my editor, Lauren Wein, who challenged and championed me equally. The book is better because of you and so am I.

Thank you to everyone at Virago Press, most of all Rose Tomaszewska for the excellent edits and support.

To Camille Morgan, thank you for your guidance, and to Elizabeth Wachtel, thank you for your joy, brilliance, and sustaining belief.

Thank you to Alison Forner for my cover design and Paul Dippolito for the interior design.

I am grateful for Linda Sawicki's copy edits, which considerably improved this book.

I could not be luckier to have found a home at Avid Reader. Thank you to everyone who made *Easy Beauty* possible: Morgan Hart, Caroline McGregor, Gil Cruz, Brigid Black, Alicia Brancato, Annie Craig, Amanda Mulholland, Elizabeth Hubbard, and Rafael Taveras.

For all your work and encouragement, thank you Jofie Ferrari-Adler, Ben Loehnen, Meredith Vilarello, Jordan Rodman, and, especially, Amy Guay.

Thank you Randy Dolnick and Don Hoffman for guiding me through the audiobook narration of *Easy Beauty*.

For early reads and feedback, thank you Justine van der Leun, Rachel Aviv, Naomi Huffman, and Travis Millard.

For publishing early bits of this book, thanks to my editors at *The Believer*, James Yeh and Camille Bromley, and to my editors at *Racquet*, David Shaftel and Caitlin Thompson.

This book was aided greatly by support from the Whiting Foundation and the George A. and Eliza Gardner Howard Foundation.

For conversation and friendship, thank you Agatha French, Meghan Rogers, John Holliday, Iris Moulton, Judd Nielsen, Jake Quilty-Dunn, Kevin Nguyen, Lincoln Michel, E. Alex Jung, Giri Nathan, Brendan Klinkenberg, Isaac Fitzgerald, Justin Runge, Daniel Rolf, Robert J. Baumann, Jess Johnson, Rebecca Evanhoe, and Cote Smith.

Thank you to the Cooper family, especially Georgeanne Cooper and Jon Baker.

And, for being the four cardinal directions on my compass, thanks most of all to Merrilee Cooper, Kate Lorenz, Andrew Grossardt, and Wolfgang Cooper Grossardt.

About the Author

Chloé Cooper Jones is a philosophy professor and journalist based in Brooklyn, New York. She is a contributing writer for *The New York Times Magazine*, a Whiting Creative Nonfiction Grant recipient, and, in 2020, was a finalist for the Pulitzer Prize in Feature Writing.

Avid Reader Press, an imprint of Simon & Schuster, is built on the idea that the most rewarding publishing has three common denominators: great books, published with intense focus, in true partnership. Thank you to the Avid Reader Press colleagues who collaborated on *Easy Beauty*, as well as to the hundreds of professionals in the Simon & Schuster audio, design, ebook, finance, human resources, legal, marketing, operations, production, sales, supply chain, subsidiary rights, and warehouse departments whose invaluable support and expertise benefit every one of our titles.

Editorial
Lauren Wein, *VP and Editorial Director*
Amy Guay, *Assistant Editor*

Jacket Design
Alison Forner, *Senior Art Director*
Clay Smith, *Senior Designer*
Sydney Newman, *Art Associate*

Marketing
Meredith Vilarello, *Associate Publisher*
Caroline McGregor, *Marketing Manager*
Katya Buresh, *Marketing and Publishing Assistant*

Production
Allison Green, *Managing Editor*
Morgan Hart, *Production Editor*
Alicia Brancato, *Production Manager*
Paul Dippolito, *Interior Text Designer*
Cait Lamborne, *Ebook Developer*

Publicity
Alexandra Primiani, *Associate Director of Publicity*
Katherine Hernández, *Publicity Assistant*

Publisher
Jofie Ferrari-Adler, *VP and Publisher*

Subsidiary Rights
Paul O'Halloran, *VP and Director of Subsidiary Rights*

EASY BEAUTY

CHLOÉ COOPER JONES

This reading group guide for Easy Beauty *includes an introduction, discussion questions, ideas for enhancing your book club, and a Q&A with author Chloé Cooper Jones. The suggested questions are intended to help your reading group find new and interesting angles and topics for your discussion. We hope that these ideas will enrich your conversation and increase your enjoyment of the book.*

Introduction

Born with sacral agenesis, a visible congenital disability that affects her stature and gait, Chloé Cooper Jones found solace in "the neutral room"—a dissociative space in her mind that offered her solace and self-protection, but also kept her isolated. When she becomes pregnant (disproving her doctor who had assumed her body was "inhospitable" for carrying a child), something necessary in her starts to crack, and she must reckon with her defensive positionality to the world and the people in it. So begins an odyssey across time and space as Chloé—while at museums, operas, concerts, sporting events, and in the presence of awe-inspiring nature—reconsiders the consciousness-shifting power of beauty. Every chapter is a moving tapestry of memory, place, and ideas confronted and overturned as Chloé offers us a chance to interrogate our history and our place in it, and open ourselves up to a new story.

Topics & Questions for Discussion

1. Flip through and note the chapter and section titles in *Easy Beauty*. Some, like "The Berninis" and "The Peter Dinklage Party," refer to art and instances in the text that are relatively concrete, while others—"Go, Thoughts, on Golden Wings" and "Above the Middle Range"—are more theoretical. Consider any title: Why might Chloé have chosen this title to encapsulate the many threads of that specific chapter or section?

2. Two inciting incidents stand out in *Easy Beauty*: the conversation in the bar in Brooklyn that begins the book, and the birth of Wolfgang. What do each of these events represent? Where in the book does Chloé return to these points in time, and why? How are they connected, and how does she reconcile herself to them?

3. Conversations with and memories of her parents shape Chloé's understanding of art, motherhood, and the life she desires. Think back to her father's blue eyes (39) and his suicide note (78); to her mother's preoccupation with chores (40) and the conversation they have on the Miami Beach boardwalk (234). What is revealed of Chloé's childhood, and of their relationships to each other? How does this impact how Chloé operates and evolves throughout the book?

4. Chapter 6, "The Weakness of the Spectator," occurs about halfway through *Easy Beauty*. It includes Chloé's description of The Beyoncé Experience and how she arrived at tickets, her childhood understanding of what her disability meant to others, and the reader's first explanation of easy versus difficult beauty. The end of this chapter marks a turning point in Chloé's journey, and signals the beginning of Part 2: "The Kestrel." What makes Chapter 6 effective in pushing Chloé further from her neutral room? What key realizations does Chloé disclose to the reader here?

5. For better or for worse, Chloé takes instruction on new ways of being and seeing from every character she meets. What did she learn from Sharon, Chetra, even the indifferent man? What did you learn? Imagine a meeting between minor characters who do not cross paths in real life. What would Judd and Peter Dinklage, or Jay and Chloé's Girl Scouts troop leader, say to each other in a radically honest conversation?

6. Think back to the art Chloé encounters and ruminates on in *Easy Beauty*. How do her reactions to Bernini's *Proserpina* (6), Marc Quinn's *Alison Lapper Pregnant* (158), and Richard Serra's *Torqued Ellipses* (246) inform our understanding of her recalibrating psyche? What about *Nabucco* (73) and Marina Abramovcić's *Rhythm 0* (113)?

7. In Chapter 3, Chloé remembers her favorite Iris Murdoch essay about the transformative power of beauty, in which the novelist sees a kestrel and "In a moment everything is altered." At the time, Chloé uses her hazy memory of the piece to justify her belief that she is better off feeling "a divine loneliness" (74); then we discover, in Chapter 9, the kestrel to Murdoch is instead a way to dissolve "the brooding self with its hurt vanity" (224). How

has Chloé's perspective changed to match the true intent of this essay?

8. Andrew exists on the periphery of much of the text, yet his steady presence anchors Chloé from afar. How does Andrew diverge from Chloé, especially in conversations about Wolfgang, and what does that demonstrate about his character? In what ways does he give Chloé what she needs in order to discover the meaning of beauty for herself?

9. Wolfgang, uncommonly sensitive, "wants to go back to before he knew about other people's minds" (217). A wise child, he inspires or initiates some of the most quietly pivotal moments in the book—his escapes at the museum (244); his ebbing cynicism toward The Great Moody Trudy (268); an autumn walk in Brooklyn (270). How does Wolfgang fit into this "endless puzzle" (224)?

10. Reflect on how your perception of *Easy Beauty* shifted or changed completely as you read. Did you begin reading with expectations about how the story would unfold? Were there moments and decisions throughout the book that surprised you? Pretend you are the protagonist in Chloé's father's unpublished children's book. What "new knowledge" do you think you have acquired (238)? Ultimately, what does Chloé learn?

Enhance Your Book Club

1. As a group, pull up a world map and identify all the places Chloé mentions and visits in *Easy Beauty*, then trace her travels according to the book's timeline. Discuss how this exercise expands your understanding of how geography influences the emotional inflection points of this memoir.

2. Have each member of your group write down five pieces of any kind of culture—the selections could include celebrities, poems, architecture, film, video games, and beyond. Combine them into one list, anonymously or not, and debate whether each is an example of easy or difficult beauty.

3. Brainstorm a list of other memoirs that deal in art criticism, disability, motherhood, travel writing, or sports journalism, and discuss how these selections differ from or are similar to *Easy Beauty*. How do style and content affect your reading? What did you appreciate about Chloé's approach?

A Conversation with Chloé Cooper Jones

Is there a story or character that you wanted to include but ended up omitting? If so, why?

I probably wrote three books' worth of material before carving out what remains in this book. There are plenty of topics I wrote about or even touch on briefly here that I'd like to explore, but this book has to be about *something* and not *everything*, so cuts had to be made! Luckily, I had the very best editor, Lauren Wein, to help me make those cuts. I also happen to have the very best agent, Claudia Ballard, who helped me focus the book before it was sold.

Was it difficult to write about the ambivalence you felt toward Andrew and Wolfgang and the future they embody?

I found it liberating. Love is such a vast and complex feeling and to truly love another person is arguably the most important human activity. I think it's both reductive and harmful to ourselves and others to talk about love's "positive" aspects only, as if love should always feel certain or good. I think it is more generous to say: Love is big! So big it encompasses and includes many other emotions, like ambivalence, fear, devotion, obligation, resentment, excitement, and on and on. These feelings are not in opposition to love, but are love's texture.

What decisions do you make when attempting to encapsulate a whole person in a few passages?

I'm not attempting to encapsulate a whole person in a few passages. Nor am I attempting to encapsulate myself in a whole book. I'm telling a focused and intentional story about the struggle between the desire to live a distanced, protected, romanticized life and a more present life that necessarily exposes me to more emotional danger. It is only necessary to include the details of my life and the lives of others that serve to further the questions that arise from that struggle. There are so many more things to be said about my parents, Andrew, my son, myself, but those things aren't in the book because they don't serve the thesis of the book.

*You write about so many kinds of beauty—*cacio e pepe *in Rome, a sublime sandstorm, Wolfgang's smile. Is there a scene, sentence, or section in* Easy Beauty *that you find especially beautiful, one that has remained with you?*

If asked to quote a sentence from the book about beauty, the one that immediately pops to mind is: "Frozen gyozas again." First of all—that's a really fun sentence to say. It's very pleasing in the mouth. But it also comes in the last chapter where I'm writing about the beauty of my everyday life—walking my son to school, the sounds of Andrew making coffee in the morning, the daily discussion of what we'll eat for dinner. Our answer here being frozen dumplings. Again.

The book starts with me in one of the most exalted museums in the world, looking at a very famous figure carved in marble by Bernini, and the book ends with me in my home, looking at the living figures there, seeking beauty there. That's all very intentional and I think few sentences capture that better than "Frozen gyozas again." The recursive sounds in that sentence—the repeated "g" and "z" sounds—mimic the recursive nature of the implications of the sentence. Most families have their classic "go-to" meals, the Tuesday night dinners, that—over the course of a lifetime—you eat in each other's company over and over. These "Tuesday night dinners" can offer a very specific look into the shared intimacy of a family. In the beginning of the book, I find it challenging to recognize the beauty of this kind of everyday intimacy, which is the foundation of familial love. I take it for granted. But by the end of the book, I'm writing about frozen dumplings with the same care and attention that I'd previously given the Bernini sculptures.

As you were developing and writing Easy Beauty, *did you turn to any other books or media that inspire you? If so, what are they and how did they influence you?*

Yes! So many. The scholarly works on disability theory and disability aesthetics done by Tobin Siebers and Rosemarie Garland-Thomson were essential. As were essays by Harriet McBryde Johnson.

Do you think it is possible to arrive at something like an equilibrium regarding solitude and connection in your life? Do you think writing Easy Beauty *played a role in approaching that?*

I think it is possible to arrive at a moment of equilibrium but the work to stay there never ends.